Introducing Stratigraphy

Other Titles in this Series:

Introducing Astronomy

Introducing Geology – A Guide to the World of Rocks (Third Edition)

Introducing Geomorphology

Introducing Metamorphism

Introducing Meteorology – A Guide to the Weather

Introducing Mineralogy

Introducing Natural Resources

Introducing Oceanography

Introducing Palaeontology – A Guide to Ancient Life

Introducing Sea Level Change

Introducing Sedimentology

Introducing Tectonics, Rock Structures and Mountain Belts

Introducing the Planets and their Moons

Introducing Volcanology – A Guide to Hot Rocks

For further details of these and other Dunedin Earth and Environmental Sciences titles see
www.dunedinacademicpress.co.uk

INTRODUCING
STRATIGRAPHY

Paul Lyle

DUNEDIN

EDINBURGH ◆ LONDON

Published in the United Kingdom by
Dunedin Academic Press Ltd
Head Office:
Hudson House, 8 Albany Street, Edinburgh, EH1 3QB
London Office:
352 Cromwell Tower, Barbican, London, EC2Y 8NB

www.dunedinacademicpress.co.uk

ISBNs
9781780460222 (Paperback)
9781780465593 (ePub)
9781780465609 (Kindle)

British Library Cataloguing in Publication data
A catalogue record for this book is available from the British Library

Typeset by Westchester Publishing Services
Printed in Poland by Hussar Books

Contents

©Mary Capriole/Shutterstock

To Beth and Jack for their edification!

Preface

This book is envisaged as a guide to the understanding of geological time by applying the fundamentals of stratigraphy to the sub-division of the geological column. It is aimed at undergraduate students learning stratigraphy for the first time, amateur geologists wishing to understand the formalities of stratigraphic nomenclature, and the general reader who wants an explanation for the distribution of different layers of rock across the landscape. In the past, stratigraphy was taught in a piecemeal fashion, with each specialist concentrating on that particular section considered to be the most important. This book attempts a unified/integrated approach to Earth history and the rocks that make up the stratigraphic column.

I write as a user of stratigraphy rather than as a specialist stratigrapher. My hope is that the reader will use the book to explore and understand the concept of geological time and see how the science of stratigraphy can be used to measure and interpret the changes that have taken place throughout Earth history. The early chapters deal with the recognition of the immensity of geological time and the development of stratigraphy as a science. The Scientific Revolution of the seventeenth century involved radical changes in how the Earth was viewed in relation to the Solar System. The change from a geocentric view of the Earth's position to a heliocentric view was at odds with the orthodoxy of the established church of the time, and scientists such as Galileo were victimized for their views. Similarly, the idea that geological time was likely to be much greater than that suggested by the biblical account was held by the church authorities for many years to be heresy. By the mid-nineteenth century, however, as a result of work by pioneers such as James Hutton in the late 18th century, the true extent of geological time was becoming more apparent and many of the periods of the stratigraphic column were in place. The next phase in the early twentieth century was the development of absolute dating methods to quantify the succession. This has culminated in the precise radiometric dating techniques that have enabled the increasingly detailed sub-division of the stratigraphic column, particularly in the Precambrian.

In chapter 5 I have endeavoured to produce an integrated account of geological history as illustrated by the structure of the stratigraphic column. Some stratigraphic boundaries are the result of catastrophic events such as asteroid impacts or prolonged periods of volcanic eruptions that left worldwide marks. Other changes, such as the evolution of the atmosphere and glacial periods, resulted in less obvious boundaries on a global scale. The protocols used by the International Commission on Stratigraphy to regulate stratigraphic nomenclature using the concepts of Global Boundary Stratotype Section and Points (GSSP) and the 'Golden Spike' are dealt with. The current International Chronostratigraphic Chart is included for reference. The final part of the book deals with the application of stratigraphic data in the modern world, its relevance in the field of environmental monitoring and the use of sequence stratigraphy as an exploration tool in the hydrocarbon industry.

Geologists aspire to write geological history, and above all this requires an understanding of the complexities of geological time. By providing a calibrated framework, stratigraphy can help in this understanding and time can be considered as a succession of events, placed in the order of their happening and with regard to their duration.

Dr Paul Lyle
C. Geol, FGS
March 2019

Acknowledgements

I am happy to acknowledge those colleagues and friends who have helped me in the production of this book. John Arthurs read and commented on the entire manuscript, more than once. His detailed analysis and constructive comments have done much to improve both the content and structure of the book, and once again I am extremely grateful to him for his help. Karen Parks willingly gave me the benefit of her considerable experience, and her insights into the teaching of stratigraphy have been invaluable. I thank her for her time and interest. My thanks for a comprehensive and helpful review go to Dr Stuart Jones of the University of Durham. His comments on the text were invaluable, particularly on the topic of sequence stratigraphy, and I greatly appreciate his input. Gary Smylie provided valuable IT help at crucial times during the project, for which I am extremely grateful. I would also like to thank Anthony Kinahan, David McLeod, Sandra Mather and Anne Morton of Dunedin Academic Press for their editorial and design skills which have contributed significantly to the final version of this book.

As usual my family, in Northern Ireland and Scotland, have supported me in many ways, both directly and indirectly, and I appreciate their encouragement. I suspect they feel that writing books keeps me usefully occupied and there are worse things I could be up to. As always, Sylvia Lyle has been a constant support, both practically in terms of her proof-reading skills, but also her encouragement throughout the whole project has been invaluable.

Notwithstanding all of this support and encouragement, any errors or omissions are solely my responsibility.

1   Introduction: what is stratigraphy?

Geology is that science that provides insight into the history of the Earth as recorded in the rocks. As such, it deals with the physical structure and composition of the Earth, its history, the origin and evolution of life and the processes forming and modifying the rocks seen at the surface. This involves interpreting the evidence for **plate tectonics** whereby continents move around the surface of the Earth and ocean basins open and close in cycles lasting hundreds of millions of years. An understanding of these changes is fundamental to the understanding of natural hazards and the exploration and exploitation of natural resources, including water. Knowledge of past climatic conditions and how they have changed over geological time is vital in mitigating the effects of current climatic changes that may be due, at least in part, to human activities such as fossil fuel consumption.

Since the earliest days of *Homo sapiens* there has been a fascination with time and the origin of the Earth. The question of how much time has been available for geological processes has always been fundamental to the science, and continues to be so today, as many people in the world cling on to the belief that the Earth is less than 10,000 years old. The subject of this book is **stratigraphy,** that branch of geology dealing with sequences of rocks and geological history. The spectacular layering shown in Figure 1.1, from the Painted Desert in Arizona, USA, is a good example of a stratigraphic sequence, representing an episode of Earth history. The science of stratigraphy can be thought of as putting things in order or ordering events in time.

One of the earliest recorded commentators on sequences of rocks was the Greek historian Herodotus (484–425 **BCE**)[1], a contemporary of Socrates and often referred to as the 'Father of History'. He made the connection between the formation of sedimentary layers and time and in his book *The Histories* he noted that river **deltas** such as the Nile in Egypt formed by the accumulation of mud and sand particles deposited each year when the river flooded. Figure 1.2 shows a satellite image

[1] BCE: before the Christian or Common Era

Figure 1.1 Rock sequence, Painted Desert, Arizona.
©Mary Capriole/Shutterstock

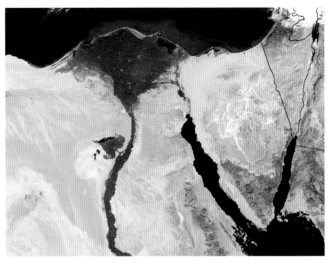

Figure 1.2 Egypt and the Sinai Peninsula with the Nile Delta and the Mediterranean Sea at the top of the image.

of Egypt and the Sinai Peninsula with the Nile delta and the Mediterranean Sea at the top of the image. The sea offshore from the mouth of the delta clearly shows lighter tones, indicating the sediment that has been carried down by the river and dumped offshore, thus building the delta out into the sea.

Herodotus recognized that it would have taken tens of thousands of years to build such a large-scale feature – an impressive intellectual leap for the age he lived in.

An example of an even more complex delta system is to be found in the Mississippi delta, forming in the Gulf of Mexico. It is estimated that around 400,000 tons of sediment are deposited in the delta each year (Fig. 1.3).

Deltas such as the Nile and the Mississippi over many thousands or millions of years build up thick sequences of sandstones and mudstones, each layer representing the sediment load for one year (Fig. 1.4). The Mullaghmore Sandstone in the west of Ireland (Fig. 1.5) is such a succession of sandstones and shales deposited in a marine deltaic environment during what is known as the Carboniferous Period around 300 million years ago (mya).

Figure 1.3 Mississippi delta showing sediment plumes.

Figure 1.5 Mullaghmore Sandstone, Co Sligo, Ireland – a sequence of sandstones laid down in a marine deltaic environment.

Figure 1.4 Deltaic deposits forming in a lake or marine environment.

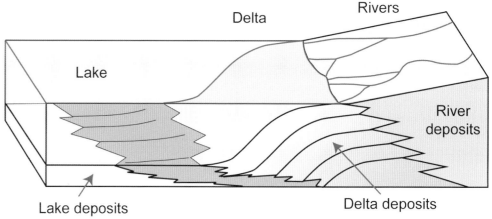

Geologists must therefore take time into account in their investigations, and an important part of geology is the study of how Earth's materials, structures, processes, and organisms have changed over time. The crucial theme running through geology as a science is the true extent of geological time. The rocks shown above are **sedimentary**, that is they are composed of fragments of pre-existing rock cemented together, or they have precipitated as mineral grains from water near the Earth's surface. Sediment is material that settles from air or water to form layers. These layers are known as **strata** (singular stratum) and each one has a distinct set of compositional or physical characteristics such that it can be readily distinguished from the layers above and below. Thus the variation in colour seen in Figure 1.1 reflects differences in composition and in the conditions under which the layers were laid down. **Sedimentary rocks** are formed as part of the **rock cycle**, along with **igneous rocks** and **metamorphic rocks** (see Fig. 1.6).

Igneous rocks are derived from molten **magma** when it cools in the **crust** or on the surface forming rocks such as **basalt** and **granite**. When subjected to weathering and erosion, the fragments of these rocks form various types of sedimentary rocks such as **sandstone** and **limestone** after they have accumulated and been compacted and **lithified**. If sedimentary or igneous rocks are buried sufficiently deeply in the crust they are changed by the increased heat and pressure to metamorphic rocks such as **gneiss** and **marble**. Metamorphic rocks can also be subjected to weathering and erosion, thus also contributing to sedimentary rocks. If subjected to further increased temperature and pressure, they may pass their melting point and form magma, thereby contributing to igneous rock production. The science of stratigraphy is primarily concerned with the distribution in time and space of sedimentary rocks, but igneous and metamorphic rocks also fulfil an important role in calibrating stratigraphic successions.

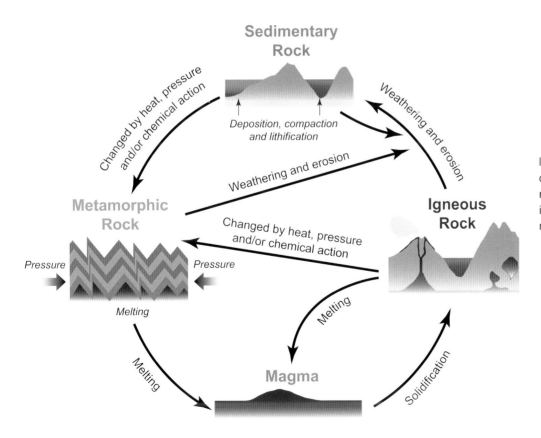

Figure 1.6 The rock cycle illustrating the relationship between igneous, sedimentary and metamorphic rocks.

Since sediments form by the accumulation of particles that have fallen down through air or water – indeed, the term comes from the Latin *sedimentum* meaning settled – it is clear that those layers at the base of the sequence were laid down before those nearer the top. An important principle in stratigraphy is therefore easily established: that sedimentary sequences such as those in the Painted Desert go from oldest at the base to youngest at the top. This is the Principle of Superposition, that when sediments are deposited, those which are deposited first will be at the bottom. In this way the connection is made between sediment accumulation and the passage of time. A second important principle that stems from the first is that the sediment layers are laid down flat – the Principle of Original Horizontality.

Varves

A particular sedimentary environment where the deposition for each year can be identified and measured is the formation of varves in a glacial lake. A varve is a pair of sedimentary layers that form in an annual cycle as the result of seasonal weather changes. They typically form in lakes found at the front of glaciers or ice caps and the two layers, or couplet, consists of a coarser-grained summer layer formed during open-water conditions, and a finer-grained winter layer formed from deposition from suspension during a period of winter ice cover. Many varve deposits contain hundreds of couplets (see Fig. 1.7).

These rhythmic deposits with alternating parallel layers can be used in stratigraphy to date events. In this respect, the

Figure 1.8 Pleistocene varved clays, Baumkirchen, Tyrol, Austria.

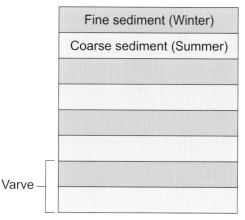

Figure 1.7 Varve couplets showing coarse summer and finer winter components.

varves are similar to tree rings. Most people are aware that by counting the rings across a tree stump the age of the tree can be calculated, since a new growth ring is added each year. It is not so well known that it is also possible to glean some information about the growth conditions that prevailed during each year. For example, warm and wet years produce rings that are thicker than average, while cold, dry years result in thinner ones (see Fig. 1.8).

Using varves, geologists have been able to build up a similar broad picture of the variation of climate over time in a particular area. In addition to the overall change from coarse to fine laid down from summer to winter, the amount of material laid down in a particular year will depend on the weather conditions that prevailed. A notably hot summer will cause more ice to melt, producing more sediment and therefore forming a relatively thick varve, while thinner varves result from cool summers. By examining the patterns of varves from different areas it is possible to recognize exceptional or distinctive seasons, known as marker horizons, and thus correlate the seasons. **Correlation** is the recognition of an event in the sequence of sediments in more than one area, showing the sediments to be the same age.

Figure 1.9 illustrates how a marker horizon, in this case a brown clay, can be used to match sections of varve clays across a number of locations. See Figure 1.9A and 1.9B. Once

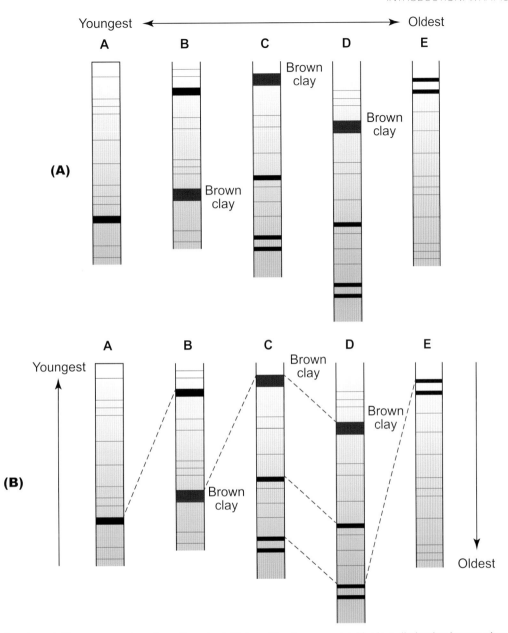

Figure 1.9 **(A)** Sections A-E are varve clays from five glacial lakes. The layer marked bc is a distinctive brown clay. Other thick black lines represent exceptional years with an abundance of organic material in the annual deposit. **(B)** Lines of correlation are shown in red. The brown clay layer can be correlated across sections B, C and D and the two thick beds at the top of E correlate with those near the bottom of C. The prominent bed near the base of section A correlates with that at the top of B. (Continued)

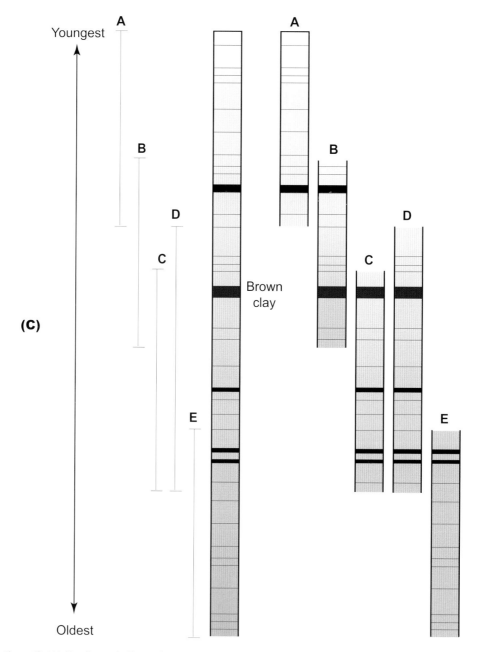

Figure 1.9 (Continued) **(C)** Sections A-E can be used to construct a complete **stratigraphic column** for the whole region by combining all the beds in order from oldest at the base to youngest at the top.

this correlation has been made, a complete sequence or **stratigraphic column** can be constructed from all five sections (see Fig. 1.9C).

The combination of Sections A–E in Figure 1.9 is used to construct a complete section for the whole region. The integration of all the beds in order of deposition from the oldest at the base to the youngest at the top is referred to as a stratigraphic column.

Lithostratigraphy is a correlation based on the composition and texture of the rocks. Other characteristics of the rocks used to produce a stratigraphic column include:

- **biostratigraphy**: the fossil content, an important aspect of stratigraphy to be dealt with in detail in later chapters;
- **chronostratigraphy**: the age of the rock irrespective of its composition or texture;

- **chemostratigraphy**: the chemical composition of the rocks;
- **magnetostratigraphy**: the magnetic properties of the rocks; and
- **sequence stratigraphy**: sedimentary units defined by unconformities.

The process of constructing the stratigraphic column has taken several hundred years, and in that time fossils and the science of palaeontology have played an important role in establishing the order of events in Earth history. This involves measuring the **relative age** of a rock, its age compared to another – it can be older, younger or the same age. In contrast, the **absolute age** of a rock is its age measured in years. The measurement of absolute age and its role in stratigraphy will be discussed in detail in chapter 4.

2   The historical development of stratigraphy

In 1666 a 29-year-old Danish-born doctor of medicine named Nils Stensen, now usually anglicized to Nicholas Steno, arrived at the Medici court in Florence. Tuscany was ruled at that period by the Grand Duke Ferdinand II, who happened to have studied physics with Galileo some years earlier, and had retained a passionate interest in scientific matters.

While Steno's early scientific endeavours were in the field of medicine, particularly anatomy, he was among the first to realise that fossils represent once-living organisms. Objects known as *glossopetrae* or tongue stones, commonly found in Malta, had many superstitions based around them, including that they either grew in the ground or dropped from the sky.

After dissecting the head of what was probably a great white shark on the orders of Ferdinand, Steno realized that tongue stones were in fact the teeth of dead sharks. This insight represented a major step forward in the recognition of the significance of fossils and their role in unravelling Earth history and understanding the sequence of life. This would be worked out in the centuries following Steno's death, but it would be done using Steno's other great contribution to science – Steno's Laws of Stratigraphy. See Figure 2.1.

Steno's Laws of Stratigraphy are: (a) the principle of superposition; (b) the principle of original horizontality; (c) the principle of cross-cutting relationships; and

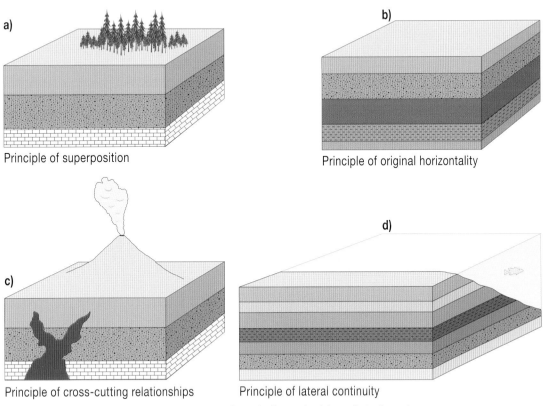

a) Principle of superposition

b) Principle of original horizontality

c) Principle of cross-cutting relationships

d) Principle of lateral continuity

Figure 2.1 Nicholas Steno's four principles of stratigraphy.

(d) the principle of lateral continuity. According to these principles:

- Younger rocks sit above older deposits (a).
- Rocks were laid down originally horizontally by the deposition of particles falling out of suspension in a fluid (b).
- The law of cross-cutting relationships states that an event that cuts across existing beds of rock must be younger than the original strata (c).
- The law of lateral continuity states that layers of rock can be assumed to have continued laterally far from where they presently end, an important factor when considering inter-regional correlations (d).
- The principle of included fragments (principle of contained fragments) states that any rock containing fragments of another rock body must be younger than the rock unit from which those fragments were derived.

The last chapter introduced the concept of lithostratigraphy – the definition of beds of rock based on their physical characteristics. Lithostratigraphy is one of three sub-divisions of stratigraphy, the other two being *biostratigraphy* – the definition of strata based on their fossil content, and *chronostratigraphy* – their definition based on the absolute age of the rock in years.

Biostratigraphy

The Law of Faunal Succession states that fossil plants and animals succeed each other in a definite order, this order representing the sequence of life forms found vertically through the rocks. As geological investigations across Europe progressed in the late eighteenth and early nineteenth centuries it became obvious that there was a consistent progression of fossils from early, primitive life forms to more complex forms as the rocks became younger.

Figure 2.2 shows the evolution of relatively advanced groups such as the fishes, reptiles and mammals. These are vertebrates, animals with a backbone that evolved at a later stage than less complex animals such as brachiopods and trilobites. These latter groups do not possess a backbone and are invertebrates. **Periods** are the sub-divisions of the stratigraphic column, from oldest to youngest, bottom to top. How this succession was established is dealt with in the remainder of this chapter. Figure 2.2 here refers to Faunal Succession.

In the late eighteenth and early nineteenth centuries a man called William Smith (1769–1839) was earning his living in England as a canal surveyor.

The canal system in Britain developed as a consequence of the Industrial Revolution, which began in Britain during the mid-eighteenth century. The road system across the country at that time was poorly developed, and canals and eventually railroads provided a reliable and economic way to transport raw materials and manufactured goods in large quantities across the country. This led to a demand for resources such as coal and iron ore, and the search for them intensified as industrialization spread. While working as a surveyor on canals such as the Somerset Coal Canal, Smith tracked rock strata across the landscape.

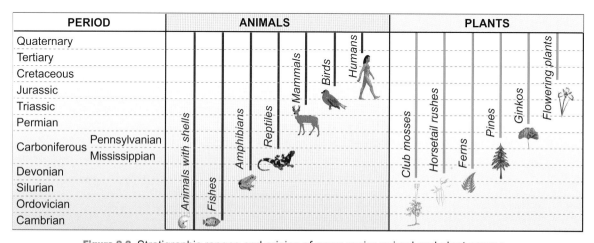

Figure 2.2 Stratigraphic ranges and origins of some major animal and plant groups.

As William Smith became more experienced, he saw that the rock strata uncovered in the excavations were found in a predictable pattern, and the same strata were always found in the same relative positions.

The Law of Faunal Succession is based on the recognition that groups of animals and plants preserved as fossils in sedimentary sequences change by evolution as time passes. When the same kinds of fossils are found in rocks from different places, then those rocks are the same age. This concept is the basis of biostratigraphy, the science of dating the rocks by using the fossils contained within them. The principal use of biostratigraphy is to establish correlations between rocks that are of the same age as determined by their fossil content (see Fig. 2.3).

It is conventional in geological mapping (when investigating the geology of an area) to group rocks into **formations**- a group of – rock strata that are horizontally continuous and with distinctive features, either of appearance or composition, that allow the formation to be recognized across a wide area. This recognition allows similar strata (see chapter 1) to be correlated with each other, and the scale of this correlation from one location to another can vary widely – from a few kilometres locally to recognition between continents. Although the two localities A and B (Fig. 2.3) could be separated by tens, hundreds or thousands of kilometres, identification of similar assemblages of fossils in a rock unit at the two localities mean that those two rock units are the same age.

By extending his fossil collecting over a wider area, Smith realized that the same groups of fossils, from older to younger rocks, could be identified in many parts of the country, and he used this information to correlate successions at the various localities he had examined. This allowed him, in 1815, to publish the first geological map encompassing England and Wales, along with part of Scotland – a truly remarkable achievement (see Fig. 2.4).

As information on the distribution of rocks and their fossils expanded in the nineteenth century, the concept of **zone fossils** or **index fossils** was developed. These were distinctive fossils that identified **biozones,** which could be used to refine the sub-divisions erected by earlier workers such as William Smith. While biostratigraphy is a relatively straightforward concept, complications arise due to factors in the biology of the preserved organisms such as their environmental range, their rates of speciation/evolution and their aptitude for preservation after death. In practice the most useful species that could be described as diagnostic are those with the fastest rate of species development and the widest geographical distribution. If a given fossil group is both wide-ranging and gives rise to rapidly evolving forms that are mineralogically strong enough to be preserved and occur in substantial numbers, then that group fulfils the requirements of an index fossil.

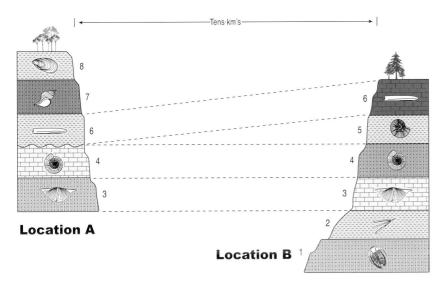

Figure 2.3 Principle of biostratigraphy.

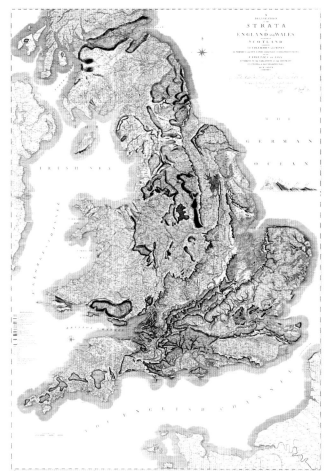

Figure 2.4 First geological map of Great Britain, published by William Smith, 1815.

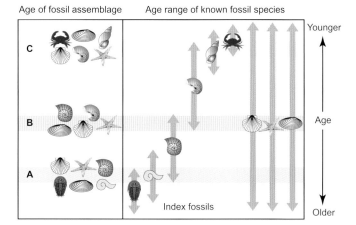

Figure 2.5 Age range of fossil species and their use as zone fossils.

Index fossils

All fossils occur over a specific time interval referred to as their range, with some species living for long periods without change in morphology, and other species having only a very short range before changing. Figure 2.5 indicates the range for a number of separate fossil species, the range shown by the double-headed arrows. The lower arrow point indicates the first occurrence of the species, the upper point the stage at which it became extinct and disappeared from the fossil record. Examination of the ranges of each of the fossil species represented shows that the three species to the right of the diagram all existed for the full extent of the sedimentary sequence and would therefore be of little use in sub-dividing the succession. In contrast, the other fossil species have much narrower age ranges, which overlap in places. For example, in assemblage B the range of the blue ammonite overlaps with the orange ammonite across a relatively narrow time zone indicated by the red box. This means that if both the blue and orange ammonites are found together in a stratum, then the rock must have been deposited during the time interval shown by the red box, the time during which the two species co-existed. This allows for a much more precise correlation than if only one or the other ammonite occurred. The blue and orange ammonites are acting as zone fossils in this instance.

In most cases the best index fossils are those that live in open water, either as **plankton** or actively swimming forms, evolving rapidly and being widely distributed. **Graptolites** are good examples of zone fossils that define biozones. They are early colonial organisms that mostly had a planktonic lifestyle, which meant they floated on the world's oceans. After death they sank to the sea bed and were fossilized (see Fig. 2.6). They were therefore widely distributed, and since they also evolved rapidly, producing many separate forms, they could be used to define relatively narrow time zones in a succession.

Graptolites fulfil many of the conditions required to be useful in stratigraphic correlation. Their planktonic lifestyle

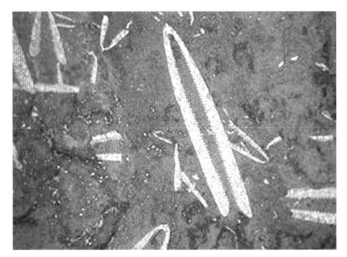

Figure 2.6 Specimens of the graptolite Didymograptus murchisoni, a zone fossil from Ordovician Period.

a stratigraphic range, in this case the Silurian Period, a time-span of around 25 million years (myr). This represents a relatively short interval in geological terms, and it means that the biozones characterized by these graptolite species are quite short, generally less than one million years.

The rocks of the Ordovician and Silurian Periods were the subject of one of the great geological controversies of the nineteenth century, involving some of the major personalities of the subject at the time. We will see later in the chapter that the eventual resolution of the dispute owed much to the unravelling of graptolite evolution.

The making/development of the geological timescale/stratigraphic column

The extent of geological time, as well as being a fascination, has from the earliest speculations been a source of acrimony and dispute. In Europe in the Middle Ages Christianity was the dominant religion. The timetable for the origin of the Earth and the age of that Earth was based on biblical accounts, largely in the Book of Genesis. This account of the Creation, coupled with a geocentric view of the universe, namely that the Sun was in orbit around the Earth, held sway until challenged by the Scientific Revolution that began in the mid-sixteenth century when European scientists set out

means they are geographically widespread and largely independent of their environment. They evolved rapidly, producing a variety of relatively short-lived forms that were capable of preservation. The forms in Figure 2.7 show examples of the huge variety of forms adopted by graptolites over

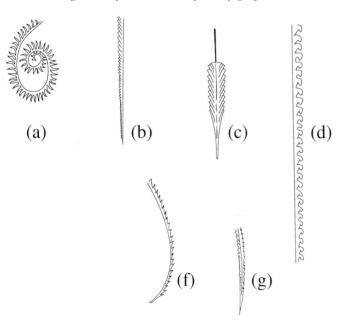

Figure 2.7 Characteristic Silurian graptolite forms.
(a) Monograptus convolutes
(b) Monograptus jaculum
(c) Cephalograptus tubulariformis
(d) Monograptus lobiferous
(e) Monograptus sedgwickii
(f) Monograptus gregarious
(g) Akidograptus acuminatus.

to change this Earth-centred view. The publication by Nicolaus Copernicus (1473–1543) of the Heliocentric model of the Solar System, which placed the Sun at the centre, was in direct contradiction to the orthodoxy of the day, and the concept was further developed by Galileo Galilei (1564–1642) and Johannes Kepler (1571–1630).

The intellectual climate of the time in Europe was largely dictated by the Church and such views were regarded as heretical and subject to disapproval by the Church authorities. One of the responses by the Church was the publication in 1650 by Archbishop James Ussher of his *Annals of the Old Testament* in which he precisely dated the date of the Creation as the night preceding the 23rd day of October, 4004 BCE. We will discuss this in a later chapter dealing with attempts to calculate the age of the Earth.

Despite the backlash from the Church about the changed perceptions of time, by the late seventeenth and early eighteenth centuries the intellectual climate was undergoing a fundamental change, and there was a growing demand to separate scientific thought from the constraints of the Old Testament. Foremost among this new wave of rationalists was James Hutton (1726–1797) from Edinburgh. He probably did more than any other individual to promulgate the idea of the vastness of geological time. Hutton was the son of a merchant who had served as City Treasurer in Edinburgh. He initially studied classics but changed to medicine and completed a medical degree in Leyden in Holland in 1749. His initial endeavours were in agriculture, running family farms in Berwickshire in Scotland, and it was here that he began to recognize the importance of the forces of weathering, erosion and sedimentation to the formation of new land areas. Returning to Edinburgh in the mid-1760s he became part of the vibrant intellectual life of that city in a period known as the Scottish Enlightenment, and turned his attention to studies of the newly emerging subject of geology.

Hutton published a paper in which he put forward his ideas on a perpetually renewing Earth, one whose development could be explained by those processes such as weathering, erosion and deposition that he had observed operating contemporaneously on his farms. He put forward the idea of a continuous cycle whereby the land is eroded and rocks and soil are washed into the sea, compacted into new sediments and then forced up to the surface by volcanic forces to be worn away again by weathering and erosion. This was an idea that came to be known as **Uniformitarianism**.

Hutton at this stage was beginning to realize how vast geological time had to be to accommodate these processes, and he used his extraordinary powers of observation to gather evidence to support his theories. His genius was to recognize the feature known as an **unconformity**.

Unconformities are gaps in the rock record. They represent a time when no deposition of sediment took place and the sequence was interrupted by a period of erosion. His most famous example is at Siccar Point in Berwickshire, east of Edinburgh in Scotland, where sloping red sandstones overlie vertical grey sandstones (see Fig. 2.8).

According to the Law of Superposition, the overlying red sandstones must be younger than the vertical grey sandstones underneath. Following the Law of Original Horizontality, both sets of beds have clearly been disturbed from their original orientation. The sequence of events is therefore:

1 The grey sandstones are deposited;
2 They are buried and folded during a mountain-building episode or **orogeny**, resulting in their present vertical alignment;
3 The mountains are uplifted and eroded to produce a near-level surface, the **plane of unconformity**;
4 This surface subsides and the second set of sediments is deposited horizontally, along the plane of the unconformity;
5 Later Earth movements tilt these younger sediments from the horizontal.

In one of the great intellectual leaps of the age, Hutton understood that, since the processes were sequential, and since each stage took a considerable time, then the total extent of geological time had to be vast. He surmised the plane of unconformity between the grey sandstones and the red sandstones represented a long period of erosion and therefore non-deposition. In fact we now know, because of modern-day dating techniques, that there is approximately 80 million years between the grey (*Silurian*) rocks and the overlying red (*Devonian*) rocks. It was at this locality in 1788 that Hutton's companion and collaborator John Playfair exclaimed in wonder- *the mind seemed to grow giddy by looking so far into the abyss of time.*

Hutton's insight into the extent of geological time and the propagation of his ideas after his death by Playfair represented a major step forward in the development of geological science at the end of the eighteenth century that paved the way for

Figure 2.8 The outcrop at Siccar Point, Berwickshire, showing sloping red sandstones overlying vertical grey sandstones.

rapid progress in the science of stratigraphy in the first half of the nineteenth century. The recognition of unconformities has been fundamental to the development of stratigraphy ever since, and unconformities provide the framework for that part of stratigraphy termed sequence stratigraphy, to be dealt with in a later chapter.

The earliest stratigraphers

Even before Hutton was speculating on the extent of geological time and the cyclical nature of Earth processes, Johann Gottlob Lehmann (1719–1767) from Saxony in Germany and Giovanni Arduino from Tuscany were taking the first steps towards the construction of what eventually became the stratigraphic column.

Lehmann was a geologist and mineralogist with an interest in ore mineralogy. In 1756 he proposed a three-fold classification of rocks based on superposition, as advocated by Steno in the seventeenth century. The oldest group he designated *Primary*, which were essentially crystalline rocks, the *Secondary* group above them were layered and fossiliferous, and the youngest group consisted of *alluvial* or river-derived unconsolidated sands and gravels. Meanwhile, a few hundred kilometres to the south in northern Italy, Giovanni Arduino (1714–1795), another mining specialist who worked mainly on the rocks of the southern Alps, proposed in 1760 a four-fold division of the rock succession. His oldest divisions, the *Primary* and *Secondary*, are roughly equivalent to those of Lehmann, but in addition Arduino proposed an additional category, the **Tertiary**. This described the poorly consolidated but still stratified and fossiliferous sediments found underneath the unconsolidated alluvial sediments that Lehmann also recognized.

Both men, one based in northern Europe, the other further south, were pioneers in the field of stratigraphy. Independently they recognized that the rocks of the Earth can be subdivided by characteristics such as composition, stratification, the presence or absence of fossils. By applying the principle of superposition, Lehmannn and Arduino were able to put the beds in order of deposition and thus establish a chronological sequence. The first steps towards erecting a worldwide stratigraphic column had been taken. The scene was now set for rapid advances in stratigraphy from the end of the eighteenth century to well into the nineteenth century, involving some of the greatest scientific names of the time and leading to the emergence of geology as a scientific discipline in its own right.

Following on from the efforts of Lehmann and Arduino, the next significant advance in stratigraphy was initiated by Alexander von Humboldt, a remarkable man whose contribution to the natural sciences over his lifetime was immense.

Alexander von Humboldt (1769–1859) was a German naturalist and explorer who did much to develop the disciplines of physical geography and biogeography after extensive travel in South America between 1799 and 1804. The Humboldt Current off the west coast of South America was named after him. Earlier in his career he developed an interest in geology and was the first person to recognize as a separate geological formation the widespread occurrence of the Jura Limestone or *Jura Kalkstein* in eastern France and then subsequently throughout Europe. On this basis he proposed the **Jurassic System** in 1795. Humboldt set off a process which, over the next 80 years or so, would lead to the erection of the stratigraphic column, which could be applied across the world and across geological time. The term **system** in stratigraphy designates rocks formed during a fundamental unit of time, the **period**.

Humboldt's recognition of the extent of the Jurassic limestone in Europe and the publication of William Smith's geological map of Britain in 1815 meant that by the early nineteenth century rock successions were beginning to be recognized on a global scale. In 1822 the Belgian geologist Jean d'Omalius coined the term **Cretaceous** (from the Latin *creta* meaning chalk) based on the deposits of chalky limestone exposed extensively in France, Belgium, the Baltic coast, and famously in the White Cliffs of Dover (Fig. 2.9).

By this time the Industrial Revolution in Britain was in full swing, and the fuel for that revolution was coal from the English Coal Measures. Thus also in 1822, two geologists from England, William D. Conybeare and William Phillips, established the **Carboniferous System** (literally *carbon-bearing*) consisting of an older sequence of mainly sandstones and limestones, the Lower Carboniferous, and a younger sequence including the coal beds, the Upper Carboniferous. With their designation of the Carboniferous, Conybeare and Phillips made the first in a series of remarkable contributions to stratigraphy by British geologists in British localities. Over the next roughly 50 years no fewer than five systems of the lower part of the stratigraphic column would be designated using localities in England and Wales. Much of this would be due to the efforts of two men, Roderick Murchison and Adam Sedgwick, who were friends and collaborators in the early part of the century but would both die in the 1870s estranged from each other after one of the bitterest scientific disputes in the history of geology.

Figure 2.9 White Cliffs of Dover.
©John Hemmings/Shutterstock

The Cambrian–Silurian controversy

Slowly and sadly we laid him down,
From the field of his fame fresh and gory;
We carved not a line, and we raised not a stone,
But left him alone with his glory.
— *The Burial of Sir John Moore after Corunna*:
    Charles Wolfe (1791–1823)

These lines from a poem about the burial of the leader of the British army at Corunna in northern Spain after a battle during the Napoleonic wars may not at first glance seem relevant to a geological dispute in the early nineteenth century. However, for one of the antagonists in that dispute it must have been a formative event in his early life before he took up the study of rocks and fossils. Roderick Impey Murchison, later to be Sir Roderick and a baronet, was born in Scotland of wealthy parents, in 1792.

After the death of his father when he was just four years old, Roderick was eventually sent to a military college in preparation for life as a soldier. In 1808 he landed in Spain with the Duke of Wellington and in 1809 at the age of 17 he took part in the retreat to Corunna and the subsequent battle with the French. These events were the precursor to a remarkable career in geology, which saw him achieve lasting fame as a stratigrapher/palaeontologist for his work in many regions of Europe and parts of Russia. After his military service he married and settled in the northeast of England, where eventually he became interested in the burgeoning science of geology, encouraged in this by his acquaintance with Sir Humphry Davy. Davy was famous for his invention of the Davy lamp for use in coal mines, but he was also an eminent chemist and Fellow of the Geological Society. This is the learned society for geology and, founded in 1807, is the oldest national geological society in the world. It played an important role in the development of geology in the nineteenth century as a distinct science by providing a forum for communication and debate of new ideas.

Murchison's career, however, was marked by often long-running disputes with colleagues concerning the establishment of the Cambrian, Silurian and Devonian periods/systems. It was obvious to his contemporaries that not only did he have a disputatious and litigious character, but he was also inclined to the use of military metaphors and to turn every scientific dispute into a military campaign. As eventual Director of the Geological Survey he ruthlessly used his position of power to advance his ideas to the detriment of any opponents. It is hard not to feel that such character traits were due, at least in part, to his schooling and early military experiences. The Murchison Medal is still one of the highest honours of the Geological Society of London.

Adam Sedgwick, the other principal protagonist in the Cambrian/Silurian controversy, was the son of an Anglican vicar, the third of seven children, who studied mathematics and theology at Cambridge, took holy orders in 1817 and somewhat surprisingly became professor of geology at the university a year later, despite having no experience of the subject beforehand. The story may be apocryphal, but on his appointment he apparently said, perhaps in his own defence *Hitherto I have never turned a stone; henceforth I will leave no stone unturned*. However, he quickly showed an aptitude for the subject and was soon making significant contributions in the fields of stratigraphy and palaeontology.

Along with Murchison, he spent much of the 1830s investigating the complex sequences of folded and faulted rocks found in Wales. These rocks were often referred to as **greywackes** (from the German *grauwacke* meaning grey stone) and consisted of sandstones and mudstones found over much of Wales but also in the Lake District of England, the Southern Uplands of Scotland, and in a wide belt in Ireland from County Down to County Longford (see Fig. 2.10).

In 1835 Murchison and Sedgwick presented a subdivision of the pre-Carboniferous rocks of Wales. Murchison in south Wales identified a distinctive set of fossils and named the rocks containing these as **Silurian**, after the *Silures,* a Celtic tribe who lived in the area in Roman times. Sedgwick, working in central and north Wales, proposed a separate sequence of rocks below Murchison's Silurian, which he named **Cambrian**, after the Latin name for Wales, *Cambria*. These proposals were contained in a joint paper in 1835, but it quickly became evident that the two systems overlapped. Since Murchison had greater fossil evidence, he soon claimed Sedgwick's Cambrian as part of the Silurian. The resulting quarrel left the two men as bitter adversaries never to be reconciled. Before this falling out, however, they had both participated in what became known as the Great Devonian Controversy involving a third colleague, Henry Thomas De la Beche (1796–1855).

Figure 2.10 Typical exposure of the greywacke series, County Down, Northern Ireland.

The controversy began in 1834 while Murchison and Sedgwick were mapping in Wales and De la Beche was mapping in nearby Devonshire. There was a general assumption that the rocks of Devon were of a similar age to those of Wales. The dispute centred on the age of fossil land plants of Carboniferous species found in coals by De la Beche in Devon, in strata that were designated as Silurian or even Cambrian. Recognition of rocks of Carboniferous age now had important economic implications with the increasing demand for coal for Britain's burgeoning industrial activities.

The older strata in Wales contained no coal or even any discernible plant life; all previously recorded coal deposits in Britain were in the much younger Carboniferous. Murchison profoundly disagreed with De la Beche's placement of these beds, insisting they were, in fact, occurring near the top of his system, the Silurian. De la Beche claimed that the absence of a distinctive rock layer, the Old Red Sandstone, meant that there was insufficient evidence to exclude these beds from the Silurian. However, fossil corals in marine beds from Torquay

in Devon were recognized by William Lonsdale (1794–1871) as being intermediate in type between those of the Silurian below and the Lower Carboniferous above. This meant that these beds were the marine equivalents of the terrestrially derived Old Red Sandstone. This was confirmed when Murchison found an inter-stratification of **Devonian** marine fossils and Old Red Sandstone fishes near St Petersburg, Russia. Further fieldwork, however, eventually showed the apparently anomalous strata in Devon were in fact Carboniferous. In 1840 Murchison found in Russia a layer similar to the Old Red Sandstone between undoubted Silurian and Carboniferous deposits. On the basis of this discovery, in 1839 Murchison and Sedgwick jointly proposed a new period, the **Devonian**. The red sandstones exposed at Hutton's Unconformity at Siccar Point (Fig. 2.8) are examples of Devonian sediments.

Ironically, this joint venture between Murchison and Sedgwick was followed relatively soon by signs of serious scientific disagreement between the two men, and by the early 1850s the relationship had deteriorated into an acrimonious and permanent estrangement. Both men died in the 1870s, Murchison in 1871 and Sedgwick in 1873, without any reconciliation of their differences. Their dispute was finally resolved by Charles Lapworth (1842–1920), an English geologist who used fossil graptolites to designate the upper part of Sedgwick's Cambrian and the lower part of Murchison's Silurian as a new system, the **Ordovician**. Lapworth's role in progressing stratigraphy was enormous – he not only resolved one of the longest-running and most acrimonious disputes in the development of geology as a separate science, but he also made a major contribution to the construction of a worldwide stratigraphic column.

The final result of the stratigraphic revisions that resulted from the Cambrian/Silurian and Devonian controversies of the mid-nineteenth century is shown below with the insertion of the Devonian and Ordovician Systems into the original Cambrian–Silurian–Carboniferous sequence.

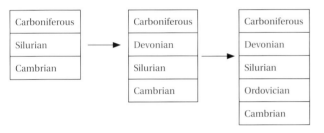

The sequence shown above is an interesting example of the way in which stratigraphy has progressed since its earliest beginnings. The three earliest systems to be designated in the Palaeozoic – Cambrian, Silurian and Carboniferous – were expanded to five systems with the addition of the Devonian and then the Ordovician as more field and fossil evidence led to a clearer understanding of the stratigraphic relationships involved.

As for Murchison and Sedgwick, with the benefit of nearly 150 years of hindsight since their deaths, we are tempted to wonder what all the fuss was about, and whether it was really necessary to destroy a close friendship in order to arrive at a conclusion that was all but inevitable in any case. The mid-nineteenth century is sometimes referred to as the 'heroic age' of geology, with larger than life characters involved in establishing many of the important principles, not only in stratigraphy but in all branches of geology. The problem with heroic ages and heroes, as we know from Greek mythology, is that they often come to unfortunate ends.

The older fossiliferous rocks up to the Carboniferous had been classified, the Jurassic had been identified since the end of the previous century, and the Cretaceous had been established in the same year as the Carboniferous, 1822. In 1834 the **Triassic** was named by the German palaeontologist, Friedrich August von Alberti (1795–1878) based on the Secondary subdivision of Lehmann discussed earlier in the chapter. Alberti proposed a division into three distinct lithostratigraphic groups – the Bunter Sandstone, the Muschelkalk Limestone and the Keuper Marls and Clays. The term *Trias* means threefold (see Fig. 2.11).

Figure 2.11 Middle Triassic marine sequence, Utah.

Sir Roderick Murchison's trip to Russia in 1840, when he was able to confirm the designation of the Devonian, was the first of two such expeditions. He returned in 1841 on a trip sponsored by the Tsar Nicholas I to try to solve a problem involving two sets of 'red beds' above and below Carboniferous sediments. Murchison's fieldwork took him to the western edge of the Ural Mountains about 1500km east of Moscow. Here he found sequences of fossils similar to those of the Silurian, Devonian and Carboniferous systems in Britain, and above them a thick sequence of sediments (sandstones, limestones, marls and conglomerates) with a fauna distinguishable from the Carboniferous below and the Triassic above. As the region was called Perm, Murchison called the new system the **Permian** (see Fig. 2.12).

Around this time, attempts were made to subdivide the geological column into groups of periods called **eras**, based on the degree of complexity of the included fossil assemblages. In 1838 Sedgwick and others proposed the term **Palaeozoic,** meaning *ancient life forms,* to include the systems Cambrian to Permian. In 1840, another English geologist, John Phillips, who incidentally was John Smith's nephew, proposed the terms **Mesozoic** and **Cenozoic**, meaning *middle life forms* and *new life forms* respectively. Taken together, these three eras are referred to as the **Phanerozoic Eon**, literally meaning the period of visible life (from the Greek *phaneros* meaning visible or evident), but now generally taken as the period of complex life forms. The preceding **eon** to the Phanerozoic is the **Proterozoic Eon,** meaning the period of earlier life, from the Greek *proteros* meaning earlier.

Designation of the Tertiary and Quaternary

After the controversies of the first half of the nineteenth century involving Murchison, Sedgwick, De la Beche *et al.*, the next major intellectual clash was that between Charles Lyell (1797–1875) and the French geologist Georges Cuvier (1769–1832). This concerned the relative roles of **U**niformitarianism and **C**atastrophism in geological history, which will be dealt with in chapter 3.

Lyell is best known as a supporter of Uniformitarianism in the middle part of the nineteenth century and as the author

Figure 2.12 Permian sandstones, Cliffs of Dawlish, Devon.
©Peter Turner Photography/Shutterstock

of the best-selling *Principles of Geology*, published between 1830 and 1833. However, by then he had already made an important contribution to stratigraphy with his proposed sub-division of the Tertiary. We have already noted that Giovanni Arduino in 1760 proposed the term Tertiary to describe the poorly consolidated but still stratified and fossiliferous sediments found under the youngest unconsolidated sediments found in northern Italy. In 1828 Lyell travelled with Roderick Murchison in France and Italy. He recognized that Arduino's term Tertiary could be applied to rocks outside Italy, particularly in the London Basin and Paris Basin. With imaginative insight he realized the youngest rocks, those near the top of the succession, contained a high percentage of fossils of living mollusc species, while those at the bottom had very few living forms, a difference he attributed to evolution. On this basis he divided the Tertiary into three parts: the **Eocene**, the **Miocene** and the **Pliocene**. These names are derived from the Greek words *eos, meion* and *pleios* meaning *dawn, less*

and *more* respectively. This refers to the beginning of modern life forms, with the proportion of modern life forms increasing in younger sediments. In the Eocene less than 10% of the molluscs were of living species, with 18% in the Miocene but more than 90% in the Pliocene. The sub-division of the Tertiary was completed in 1854 when two German scientists, Heinrich Beyrich and Wilhelm Schimper, proposed the names **Oligocene** and **Palaeocene** from the Greek *oligous* meaning *few* and *palaios* meaning *old*. Figure 2.13 below gives the complete sub-division of the Tertiary.

In modern stratigraphic terminology the Tertiary has been replaced by the **Palaeogene** and **Neogene** Periods. The Palaeogene is the lower portion of the Tertiary from the Palaeocene to the Oligocene, while the Neogene comprises the Miocene and Pliocene. See Figure 2.14.

Figure 2.14 also shows the youngest period of the stratigraphic column and the final period of the Cenozoic Era, the **Quaternary**. Proposed in 1829 by Frenchman Jules Desnoyers

ERA	PERIOD	EPOCH	Start date (mya)
CENOZOIC	Quaternary	Holocene	(0.01)
		Pleistocene	(1.6)
	Tertiary	Pliocene	(5)
		Miocene	(23)
		Oligocene	(35)
		Eocene	(56)
		Palaeocene	(65)

Figure 2.13 Subdivisions of the Tertiary.

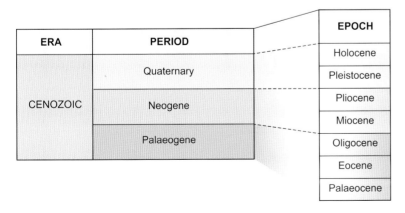

ERA	PERIOD	EPOCH
CENOZOIC	Quaternary	Holocene
		Pleistocene
	Neogene	Pliocene
		Miocene
	Palaeogene	Oligocene
		Eocene
		Palaeocene

Figure 2.14 The subdivision of the Quaternary, the youngest period of the Cenozoic.

to cover the post-Tertiary rocks of the Paris Basin, it now comprises the glacial epoch the **Pleistocene** and the epoch we are currently living through, the **Holocene**. Holocene means 'entirely recent' and includes the origin and development of the human species worldwide.

By 1840 the stratigraphic column was beginning to resemble the one we are familiar with today (see Fig. 2.15). It is important to recognize that at this stage it was entirely uncalibrated – the beds had been placed according to relative age only; there was as yet no attempt made to calculate an absolute age in years for each of these periods. That would come later.

The Carboniferous Period proposed by Conybeare and Phillips had been subdivided in the USA into *Mississippian* for the lower section, mainly limestones, with the upper section containing the coal-bearing sequences termed the *Pennsylvanian*. More recently it has been decided to use the terms Mississippian and Pennsylvanian globally.

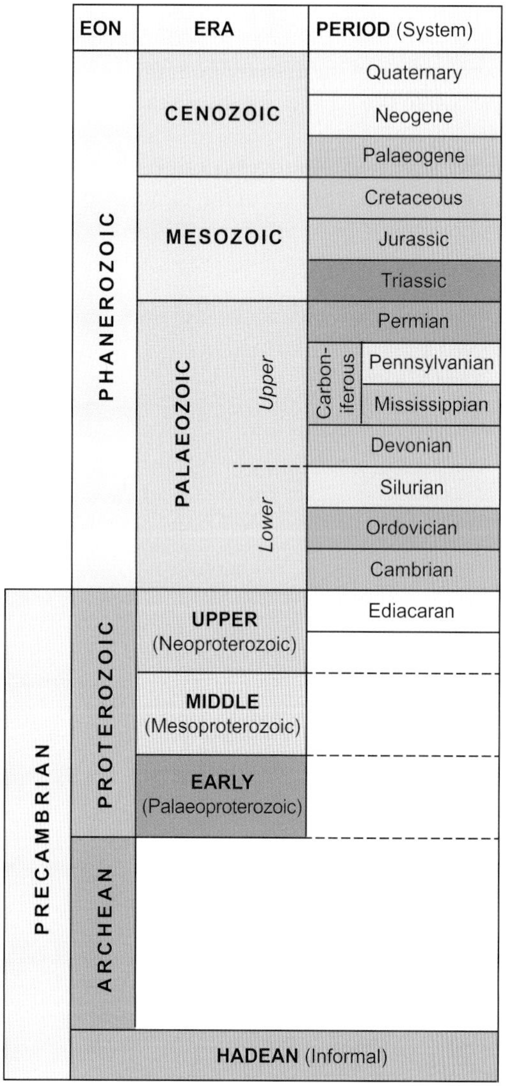

EON	ERA	PERIOD (System)	
PHANEROZOIC	CENOZOIC		Quaternary
			Neogene
			Palaeogene
	MESOZOIC		Cretaceous
			Jurassic
			Triassic
	PALAEOZOIC	Upper	Permian
			Carboniferous: Pennsylvanian
			Carboniferous: Mississippian
			Devonian
		Lower	Silurian
			Ordovician
			Cambrian
PRECAMBRIAN	PROTEROZOIC	UPPER (Neoproterozoic)	Ediacaran
		MIDDLE (Mesoproterozoic)	
		EARLY (Palaeoproterozoic)	
	ARCHEAN		
	HADEAN (Informal)		

Figure 2.15 Stratigraphical column showing the main subdivisions and periods of geological time.

3 Time's Arrow and Time's Cycle: Uniformitarianism and Catastrophism in stratigraphy

By around 1840 the main components of the stratigraphic column had been put in place. The so-called *heroic age* of geology was in full swing, and following the long wrangle between Murchison and Sedgwick over the Cambrian and Silurian the scene was now set for the next clash of geological titans – that of Charles Lyell against the French scientist Georges Cuvier, which could perhaps be viewed as a re-run of the Napoleonic Wars of Murchison's youth.

In the early years of the nineteenth century James Hutton's companion at Siccar Point in Scotland, John Playfair, published his book entitled *Illustrations of the Huttonian Theory of the Earth,* which was largely responsible for promoting Hutton's ideas after his death in 1797. Hutton had realized that the Earth had a long history and that this history could be interpreted by understanding the processes that were currently active on the Earth's surface, which was constantly being renewed by cycles of deposition and erosion. He summarized this by saying that *the future will resemble the past.* These ideas were taken up by Charles Lyell, who was destined to become one of the most influential geologists of the Victorian age in Britain. Much of this influence stemmed from his highly successful book, *Principles of Geology*, published in three volumes in 1830–33, which had as its wordy subtitle *An attempt to explain the former changes of the Earth's surface by reference to causes now in operation*. Lyell had embraced Hutton's ideas as expounded by Playfair, and so believed in the antiquity of an earth that had been shaped by the same geological forces as those currently observable, operating over very long periods of time. This view had been labelled Uniformitarianism by William Whewell, best described as an English polymath, who had a gift for words and is remembered for his first use of other terms such as *scientist, physicist* and *anode/cathode*. Lyell's views were directly opposed to those of the French scientist Georges Cuvier, a hugely respected scientist of the period. Cuvier is regarded as the founder of vertebrate palaeontology and is given the credit for establishing extinction in the fossil record as a fact. He proposed that new species were created after periodic natural events such as catastrophic floods and mountain-building episodes. This belief led Whewell to coin the term **Catastrophism** to describe the processes involved, and so the main scientific controversy of the mid-nineteenth century became Uniformitarianism versus Catastrophism, Lyell versus Cuvier.

Cuvier was the leading proponent of Catastrophism in the early decades of the nineteenth century. In a study of the terrestrial vertebrate fauna of the Paris Basin he recognized that many fossils had no living counterparts and thus appeared to be forms that had become extinct. These extinctions led to the appearance of more advanced but related forms. The concept of extinction of species was considered impossible by the religious authorities of the day on the grounds that God would not destroy His own creations. As well as religious objections, Cuvier also faced opposition for his extinction theories from Lyell, and also from Charles Darwin, who believed that species undergo gradual alteration rather than catastrophic change. The recognition today of **mass extinction events** that appear to have been triggered by events as diverse as asteroid impact, prolonged volcanic episodes or rapid changes in sea level at various times throughout the geological record is a vindication of Cuvier's pioneering views on the causes of species dying out. The significance of mass extinction events in stratigraphy will be further discussed (see chapter 5).

Cuvier believed that flooding could be a factor in the extinction and renewal of species. It was unfortunate that the expression of this view coincided with a reaction among British scientists against the idea of a Creationist origin for the Earth and a dependence on miraculous origins for the

Earth's features. Much intellectual energy had been expended on trying to fit geological observations into a framework that incorporated a worldwide flood or inundation. Cuvier scrupulously avoided all reference to religion in his scientific writing and did not subscribe to a biblical timetable for the age or origin of the Earth. His untimely death during a cholera epidemic in Paris in 1832 meant that Lyell was able to expound his theories without opposition from his principal opponent. Lyell in many ways adopted an extreme view of Uniformitarianism in that he argued that all past events could be explained by the action of processes now in operation. He refused to concede that some processes become redundant or that there could be increases or decreases of rates of development with time. It says much for Lyell's standing in the geological community that this view prevailed in certain circles for the next century and a half, to the detriment of the progression of the subject, as we will see later in this chapter when discussing the interpretation of the Spokane Flood in the western USA.

The Arrow of Time or Time's Arrow is a concept introduced by the British astronomer Sir Arthur Eddington (1882–1944) in 1927 to account for the 'one-way direction' of time. He connected Time's Arrow to the one-way direction of increasing entropy (a measure of disorder) required by the Second Law of Thermodynamics. As a system advances through time it will become more disordered and this asymmetry can be used as an empirical distinction between future time and past time. In his later work on the topic he identified a number of other 'arrows' including a cosmological arrow to explain the direction in which the universe is expanding.

In the 1980s an American palaeontologist, Stephen Jay Gould, coupled time's arrow with time's cycle and introduced them into stratigraphy. Along with the term **deep time** used by John McPhee in his book *Basin and Range* to explain the concept of the vastness of geological time, these were attempts to understand the passage of time in a geological context. With time's arrow, history is a series of linked events moving in one direction. In contrast, according to time's cycle, events have no meaning as distinct episodes and are simply parts of repeating cycles. Time in this case therefore has no direction. If we take the Bible as an example, history there is recorded primarily as time's arrow – progression was from the Creation to Noah's Flood to the delivery of the Commandments to Moses at a particular time and place. All of these events were unique and followed on from each other. However, in the Book of Ecclesiastes it says *To everything there is a season, and a time to every purpose under the heaven* (King James Bible). This represents an overlay of time's cycle on the sequence; seasons are clearly recurring events.

In geological terms non-repeating events in Earth's history, examples of time's arrow, are phenomena such as:
- the change from an early oxygen-poor atmosphere to one that is oxygen-rich;
- a greater frequency of large meteorite impacts in the early history of the Earth;
- the evolution and extinction of organisms.

Time's cycle in geology is reflected in the two cycles of deposition, erosion, uplift and deposition recognized by Hutton at the unconformity at Siccar Point (see Fig. 2.8) and other localities. The emphasis here is on periodic renewal and a uniformity of process.

Time appears to encompass distinct and irreversible events such as the impact on Earth of a large asteroid (Catastrophism), while simultaneously representing cycles of erosion and renewal and timeless order (Uniformitarianism). To fully understand the concept of geological or deep time, use of both the metaphors – time's arrow and time's cycle – is required.

Earth's cycles

The rock cycle was described in chapter 1 (see Fig. 1.6) as a fundamental process in Earth history whereby rocks are formed or changed, eroded, transported and deposited and therefore re-cycled. The stages of the cycle are readily observable; clear examples of igneous, metamorphic and sedimentary rocks are visible at or near the Earth's surface. The rock cycle is just one of a number of cyclical processes that affect the Earth. Some cycles occur on a longer timescale and can only be recognized by examining geological processes over tens or hundreds of millions of years.

The plate-tectonic cycle

Plate tectonics is one such longer-term cycle. The Earth can be subdivided into concentric layers: the inner and outer core, the **mantle** and the outermost layer, the crust. The core is about two-thirds the size of the Moon and is mostly nickel-iron in composition. The inner core is solid at a temperature of around 5700°C, surrounded by a liquid outer core that is about 2000km thick (see Fig. 3.1). The mantle is a plastic solid

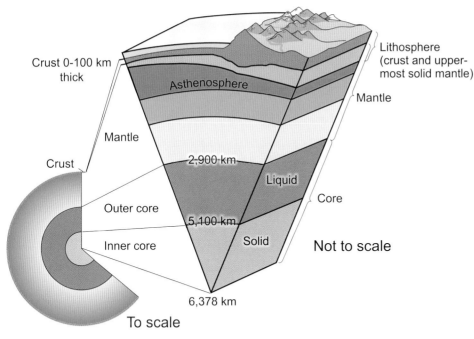

Figure 3.1 Inner structure of the Earth.
©Petrer Hermes/Shutterstock

capable of sustaining convection currents. Convection occurs in liquids when heated; the hotter material becomes lighter and rises. On cooling it becomes heavier and sinks, to be re-heated, thus setting up a circulation in the liquid. The outer layer of the mantle and the oceanic and continental crust are brittle and together make up the **lithosphere**. Directly below the lithosphere is the **asthenosphere**, a region of the mantle that is partially melted and therefore in a weaker physical state than the lithosphere directly above it.

The theory of plate tectonics states that the surface of the Earth is not fixed, but consists of a number of lithosphere **plates**, polygonal in shape, resembling the fragments of cracked egg-shell on a hard-boiled egg (see Fig. 3.2). These rigid plates move on this underlying weaker layer in the mantle, the asthenosphere.

Slow convection currents originating deep within the mantle bring hot material to the surface (Fig. 3.3). This up-welling of relatively hot material is balanced by cooler, denser material sinking back into the mantle and thus completing the convection cycle. New crustal material is generated by divergence at the mid-ocean ridges called con-structive (divergent) margins. The oceans widen by a process called **sea-floor spreading** with the new crust being generated

from the underlying mantle, and this expansion is balanced by destruction at **destructive (convergent) margins** where the plates converge and collide (see Fig. 3.4). This is a cyclical process – crust is created at constructive plate margins such as the mid-Atlantic Ridge and destroyed in collision zones such as the Andes in South America, where crust is pushed back into the mantle, or **subducted**, and is thus absorbed back into the mantle. Where continental plates are forced together the crust is lifted up to form major mountain ranges such as the Himalayas (see Figs 3.5 and 3.6).

The Himalayas were formed when India began to move northwards relatively rapidly at about 15cm per year and collided with Asia around 60 million years ago (mya) (Fig. 3.6). The ocean floor between India and Asia was systematically destroyed as the two continental plates converged and eventually collided. This movement continues, with the result that the Tibetan Plateau to the north of India is still rising at the rate of a few millimetres a year.

This mode of origin for the Himalayas accounts for the fact that the summit of Everest, the highest point on Earth (Fig. 3.7), is composed of marine limestone that formed millions of years before in a warm, clear ocean that has now disappeared. This ocean, named Palaeotethys, opened in the late

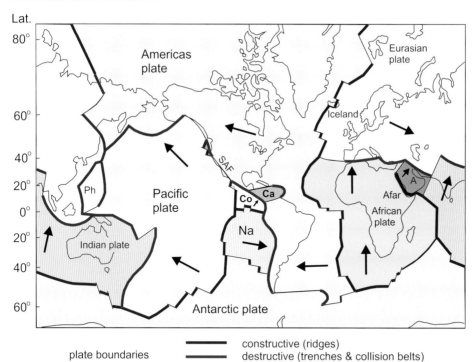

Figure 3.2 Distribution of tectonic plates on the Earth's surface.

plate boundaries

constructive (ridges)
destructive (trenches & collision belts)
conservative (faults)

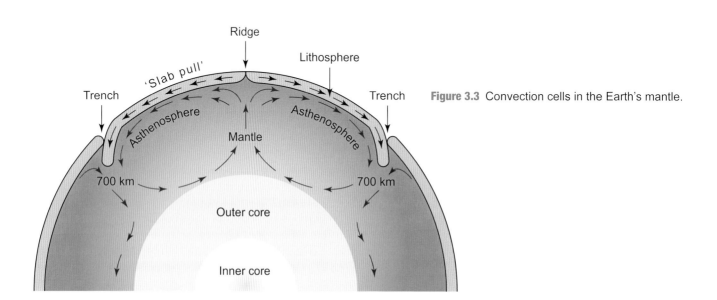

Figure 3.3 Convection cells in the Earth's mantle.

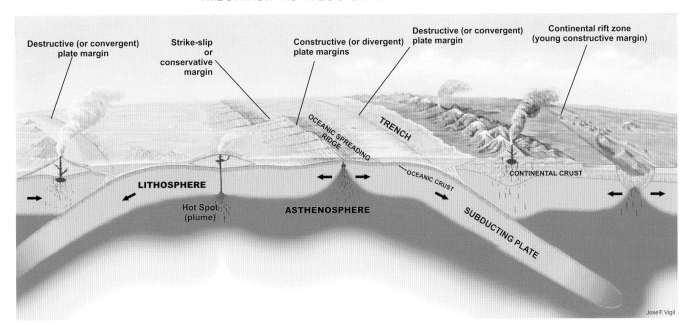

Figure 3.4 Types of plate boundaries.

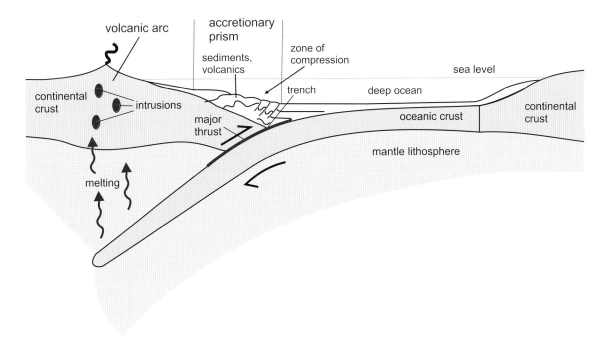

Figure 3.5 Destructive plate margin of continent-continent type.

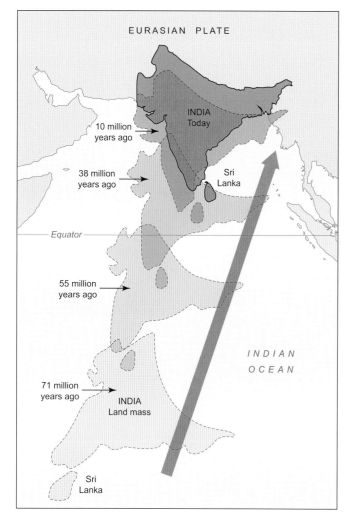

Figure 3.6 Formation of the Himalayas by the collision of India with Asia.

Figure 3.7 Mount Everest from the Nepalese Himalayas.
©Daniel Prudek/Shutterstock

Cambrian and closed in the late Triassic. It was the precursor to the Tethys Ocean, which in later Jurassic times separated the northern landmass of Laurasia from the southern landmass Gondwana.

The Wilson Cycle

The longest of Earth's cycles examined so far, the Wilson Cycle is named after the Canadian geologist John Tuzo Wilson. He realized that if ocean basins are forming by the splitting apart of continents, then somewhere else on the Earth's surface other oceans must be closing and continents colliding to form mountain ranges (see Fig. 3.8).

The Wilson Cycle, one of the largest-scale cycles operating on Earth, consists of six stages.

1 The embryonic stage: a previously stable area of continental crust is split by rising convection currents in the mantle, causing the crust to stretch and rift (e.g. the East African Rift).

2 The juvenile phase: sea-floor spreading starts and a juvenile ocean opens (e.g. the Red Sea–Gulf of Aden area).

3 The mature stage: a mature ocean basin develops (e.g. the Atlantic Ocean).

4 The declining phase: progressive closure of ocean basins occurs by subduction of ocean lithosphere after the ocean basin reaches its maximum width (e.g. the Pacific Ocean).

5 The terminal phase: shrinking of the ocean basin occurs and continental collision begins (e.g. the Mediterranean area).

6 The suturing phase: with the complete closure of the ocean basin the two continental margins of the ocean are joined or sutured together. They continue to collide and form an uplifted mountain range (e.g. the collision of India with Asia and the formation of the Himalayas and the Tibetan Plateau).

Recognition of the various stages of the Wilson Cycle is important in unravelling the complex stratigraphy in the

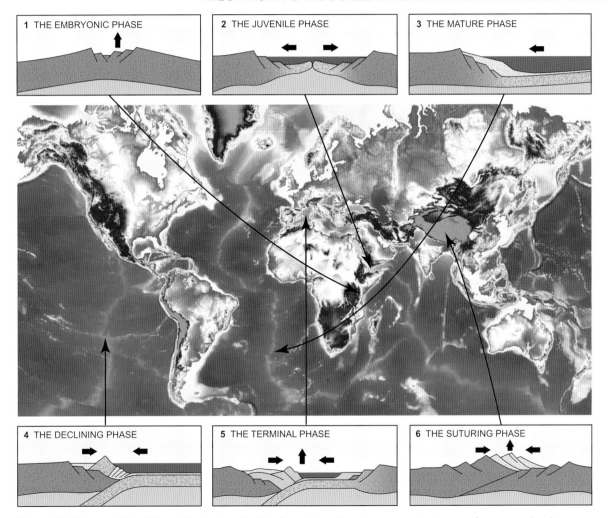

Figure 3.8 Stages of the Wilson Cycle showing the formation, development and closure of an ocean basin.

Cambrian, Ordovician and Silurian of Wales as detailed in the Murchison–Sedgwick controversy in an earlier chapter. In this case the Iapetus Ocean began to open in Cambrian times, possibly around 600mya, and by 420mya had completely closed, leaving as a legacy the complex stratigraphy of the Caledonian mountain chain, which extended from Scandinavia through Britain and Ireland to the east coast of North America at Newfoundland and down as far as the Appalachian Mountains and North and South Carolina (see Fig. 3.9).

The Wilson Cycle operates over a timescale of a few hundred million years, and an extension of this cycle is the formation of a **supercontinent**. Following the closure of Iapetus the continental areas of the Earth eventually amalgamated, so that by Permian times about 250mya the supercontinent **Pangaea** had formed. Pangaea followed a supercontinent **Rodinia**, thought to have formed about one billion years ago, and which started to break up about 600mya. The stratigraphic implications of the formation and breakup of Pangaea will be discussed in

30

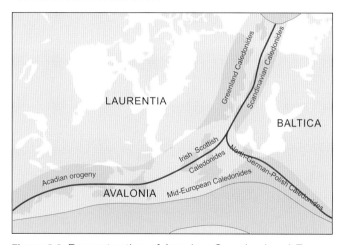

Figure 3.9 Reconstruction of America, Greenland and Europe after the closure of the Iapetus Ocean and the formation of the Caledonian mountain chain around 420mya.

Figure 3.10 Site of the Chicxulub impact crater. Gulf of Mexico. NASA image.

chapter 5. The next supercontinent is estimated to form about 250myr from now.

The reconciliation of Uniformitarianism and Catastrophism

Over time, the hard-line approach to Uniformitarianism has softened among geologists so that most would now subscribe to a doctrine of 'Catastrophic Uniformitarianism', a modification of Uniformitarianism to take into account those low-frequency/high impact events that are now known to be part of the stratigraphic record. The well-documented occurrences of mass-extinction that have taken place for a variety of reasons throughout geological history are examples of such events. Meteorite impacts on Earth such as the one at Chicxulub in Mexico (Fig. 3.10) and a more recent event such as Meteor Crater in Arizona are clear evidence of the role played by Catastrophism in the stratigraphic record. The aftermath of the Chicxulub event was probably at least a factor in the extinction of the dinosaurs at the end of the Cretaceous Period.

The Tunguska impact in Siberia in 1908 (Fig. 3.11) and the recently observed (1994) collision of Comet P Shoemaker/Levy with the planet Jupiter have alerted us to the fact that catastrophic events have a continuing role in Earth history and will therefore appear in the stratigraphic column in the future.

Figure 3.11 Trees felled at Tunguska, 1908. NASA image.

The well-known English stratigrapher Derek Ager, in his book *The Nature of the Stratigraphic Record* (1973), has summed up the situation succinctly with his comment: 'Sedimentation in the past has often been very rapid indeed and very spasmodic. This may be called the Phenomenon of the *Catastrophic* Nature of the Stratigraphical Record' (my emphasis).

Lyell's refusal to accept a role for catastrophic events influenced geological thought for many years after his death, extending to well into the twentieth century. A case in point is the Spokane Flood. In the 1920s an American geologist J Harlen Bretz proposed that the landscape in the western USA known as the 'Channeled Scablands' was the result of catastrophic flooding caused by the release of water held behind an ice dam on the Clark Fork River, some 200km to the east in Idaho. Lake Missoula, as it was known, was comparable in size and volume to one of the present-day Great Lakes.

The sudden release of this volume of water caused a flood of cataclysmic proportions, with flow velocities of greater than 80km per hour and a flow rate at its maximum estimated by the US Geological Survey as about ten times the current combined flow of all the world's rivers. The result was probably the biggest waterfall ever to exist on the surface of the Earth, Dry Falls, Washington State (Fig. 3.12). The water stripped away the glacially derived soil of the Columbia Plateau to leave a complex area of rock-cut channels and bare rock – the Channeled Scabland (Fig. 3.13).

31

Figure 3.12 Dry Falls, Washington State, USA.
©Laszlo Dobos/Shutterstock

Figure 3.13 Scabland topography, Washington State, USA.

Other evidence of water transport of sediment included giant ripple marks consisting of long gravel ridges and large blocks of rock, known as erratics, torn up and transported long distances before being dumped as the water velocity fell.

Despite such evidence, it was the middle 1960s before it could be said that the proposal by Bretz was finally accepted by the geological community, such was the legacy of the influence of Lyell from the mid-nineteenth century. Catastrophism was at last respectable, and now something akin to 'Catastrophic Uniformitarianism' was the order of the day. Features associated with both phenomena are required to fully explain the changes observed in the stratigraphic column. There is clear evidence for single catastrophic events such as meteorite impacts, just as recurring episodes such as the Wilson Cycle of ocean opening and closing indicates a degree of uniformity and repetition of process.

Time's Cycle and Time's Arrow in the Grand Canyon

The Grand Canyon of the Colorado River in the western USA is one of the geological wonders of the world. It is 29km wide and 1800m deep and displays basement metamorphic rocks approximately 1500myr old, overlain by sedimentary rocks that range from Precambrian to Permian in age, around 250myr old. This means that there is an age range of about 1250myr exposed in this one section of the Earth's crust – that is 1250myr out of the total age of the Earth of 4500myr. The Grand Canyon represents more than a quarter of the whole of geological time. It provides an ideal setting to illustrate the twin concepts of Time's Arrow and Time's Cycle (see Fig. 3.14).

Figure 3.14 Grand Canyon of the Colorado River.
©Gert Hochmuth/Shutterstock

An important feature of this geological succession is the presence of a number of unconformities – the feature recognized by James Hutton at Siccar Point in Scotland and described in chapter 2, Figure 2.11. The principal unconformities of the Grand Canyon are shown in Figure 3.15.

Unconformities not only record intervals of erosion in the rock succession but also provide a record of ancient

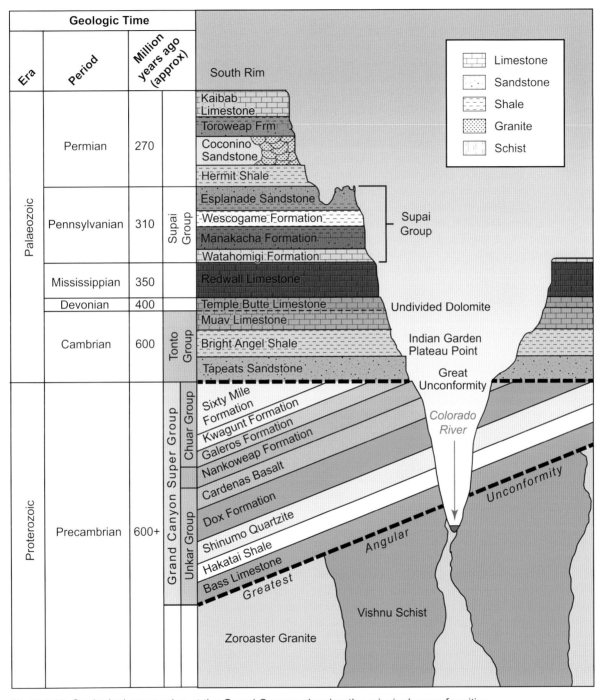

Figure 3.15 Geological succession at the Grand Canyon showing the principal unconformities.

earth movements. The oldest rocks of the Grand Canyon succession are found at the bottom of the canyon. They are in the Vishnu Formation and are metamorphic rocks intruded by granites. Immediately above the Vishnu is a group of sediments, sandstones and limestones, laid down horizontally but then tilted by subsequent earth movements. These are around 600myr old compared with the 1500myr old Vishnu Group, so the boundary between the two groups is marked, not only by a big difference in their ages, but also by a difference in their orientation. Like Siccar Point in Scotland, this is an **angular unconformity** and named the Great Unconformity by the first geologist from the US Geological Survey to map it.

Above this unconformity it is possible to recognize a further five examples, all representing crustal uplift creating new exposures of rock, which are then subjected to the processes of erosion leading to the deposition of new sediments and so continuing the cycle. As well as gaps in time, these unconformities also mark changes in the environment of deposition of the rocks – changes often associated with variations in sea level. Obvious differences in the various layers that make up the canyon walls show up as steep-sided cliffs in parts of the wall, while other parts are less steep and consist of areas of loose material or screes that cover the bare rock underneath (Fig. 3.13). The steep cliffs tend to be formed of harder rocks such as sandstone or limestone, while the loose material between is made up of softer shales and mudstones.

This variety of rock types is an indicator of the range of sedimentary environments involved. The sandstones may have formed in desert dunes or been laid down in water in a marine or river or lake environment, while the limestones represent shallow marine conditions.

The unconformities so easily seen in the Grand Canyon sequence are clear evidence of Time's Cycle with its emphasis on periodic renewal and cyclicity, but clear evidence also exists for Time's Arrow. It is possible to extract fossils from the sedimentary rocks of the canyon walls. In the oldest sediments the dominant life forms are invertebrates, characterized by trilobites. Younger beds show a progression to vertebrates with the arrival of the earliest fishes followed by amphibians and reptiles as life began to colonize the land. This progression illustrates evolution – an irreversible process with time acting in a straight line – Time's Arrow.

The facies concept and Walther's Law

Lithostratigraphic correlation is based on the composition and texture of the rocks, in contrast for example with biostratigraphy which is based on the fossil content of the rock. Much of our discussion of the historical development of stratigraphy in the previous chapter was based on the principles of biostratigraphy. We must now turn our attention to further aspects of lithostratigraphy and its role in the stratigraphic column.

The variety of rocks found in the Grand Canyon results from deposition across a range of sedimentary environments. The changes in deposition are often related to the rise or fall of sea level, for example, producing sandstones, mudstones or limestones as water levels deepen. Such changes are often slow, and this is reflected in the gradual changes seen from one rock type to another as the sea encroaches or retreats. Someone who recognized the difference between rocks defined by their composition and those defined by their age was the German geologist and mineralogist Abraham Gottlob Werner (1749–1817).

Werner is best remembered for his championing of the idea of a universal ocean that had covered the entire surface of the globe. As it gradually receded to its present level it precipitated all the rocks and minerals of the Earth's crust. The association of his theories with the universal ocean led to his supporters being dubbed 'Neptunists' after the Roman god of the sea. These theories of Werner were challenged by James Hutton in Edinburgh, who had recognized that veins of granitic rock in Glen Tilt in the Scottish Highlands must have been injected in the molten state. The recognition of the molten origin of rocks such as basalt and granite meant that Hutton and his supporters were termed 'Vulcanists' (Vulcan was the Roman god of fire). As more and more evidence concerning the origins of igneous and metamorphic rocks became available, it became increasingly difficult to accept Werner's universal ocean model, and the Neptunist model for the origin of rocks passed into the annals of the history of geology. However, by recognizing the significance of chronostratigraphy, Werner made a major contribution to the interpretation of the stratigraphic record.

For example, in chronostratigraphy a particular rock type traced across the landscape may have been deposited at different times as conditions changed across a sedimentary basin. If the sea level is rising, the beach deposit formed at the boundary of the sea and the land will become progressively younger as the sea covers more of the land surface.

The term **facies** was first used by the Swiss geologist Amanz Gressly in 1838 to describe a distinctive rock type broadly corresponding to a certain mode of origin. A facies is a body of rock with specific characteristics such as overall appearance, texture, composition or a local depositional environment. Any rock unit can pass laterally into a different facies, and such a lateral change is termed a facies change (see Fig. 3.16). As laterally adjacent sedimentary environments shift back and forth through time as a result of sea-level changes, then sedimentary boundaries also shift back and forth.

Sea-level changes occur for a variety of reasons. Glacial periods result in changes due to the enormous volumes of sea water stored in ice caps and glaciers, but also due to crustal downwarp caused by the weight of ice deforming the underlying mantle by a process called **isostatic readjustment**. Changes in the distribution of tectonic plates, as described earlier in this chapter, can alter the shape of ocean basins over long periods and thus change sea level. The causes of sea-level change will be discussed in further detail in chapter 5.

A sea-level rise is called a **transgression**. The effect of this is to produce a vertical sequence of rocks that are the result of *deeper* water environments.

The term *onlap* is used to describe the process of progressively younger beds extending onto the older rocks landwards as the transgression proceeds.

A fall in sea level is called a **regression**. In this case the effect is to produce a vertical sequence of rocks that are the result of shallower water environments.

The term *offlap* is used to describe the process of progressively younger beds being deposited seawards as the regression proceeds.

At any one time, sediments of various types are being deposited in different places. Sand is deposited on the beach, silt is deposited offshore, clay is deposited in deeper water, and carbonate sediment is deposited far from shore (or where there is little or no input of terrigenous sediment). Sedimentary environments move as sea level changes, or as a basin fills with sediment.

We can now see that the variations in rock types visible in the walls of the Grand Canyon (Fig. 3.14) are due to changes in the

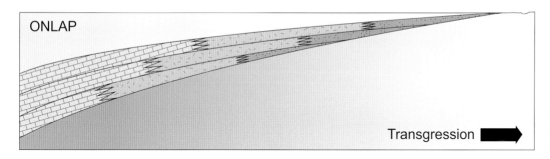

Figure 3.16 Shifting sedimentary facies in the case of sea level rise (transgression) and sea retreating (regression) After Woudloper.

environment of deposition, and as such they are facies changes resulting from transgressions and regressions. A similar interpretation is possible of the sequence shown in Figure 3.17, in the Book Cliffs, eastern Utah, USA. Here the clearly defined changes in lithologies in the cliff face reflect changes in the environment of deposition, or palaeo-environment.

Sedimentary sequences such as those above result from various lithologies being deposited in different sedimentary environments over time. This process eventually produces a vertical sequence of different facies. As laterally adjacent environments of deposition move back and forth through time as a result of transgressions or regressions of sea level, then facies boundaries also shift back and forth. With time, facies that were once laterally adjacent will shift so that the sediments of one environment of deposition will come to lie over those of an adjacent environment. This is Walther's Law, formulated by Johannes Walther in 1894. In an unbroken (*conformable*) sequence of sediments those rocks in vertical sequence represent the deposits of laterally adjacent sedimentary environments. See Figure 3.18.

The principle of Walther's Law is that the vertical sequence of sedimentary facies mirrors the original lateral distribution

Figure 3.17 Book Cliffs in eastern Utah.
©Marekuliasz/Shutterstock

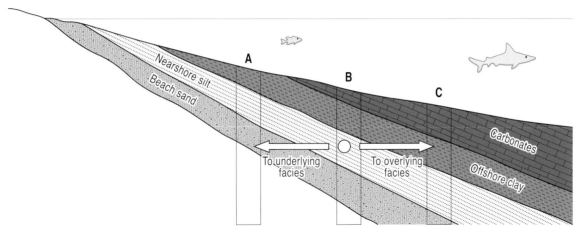

Figure 3.18 Illustration of Walther's Law of Correlation of Facies.

of sedimentary environments. When a depositional environment moves laterally as a result of transgression or regression, the vertical succession and lateral sequence of facies will be the same. Walther's Law provides a means of subdividing the stratigraphic record in response to relative rates of change of sea level. It forms the basis of a branch of stratigraphy known as sequence stratigraphy and it is used extensively in the oil industry to identify and predict the occurrence of facies as an aid to correlation of geographically separate sedimentary successions. This topic will be dealt with in detail in a later chapter.

The quantification of the stratigraphic column: absolute time

Absolute, true and mathematical time, of itself, and from its own nature flows equably without regard to anything external, and by another name is called duration:

—Isaac Newton (1642–1727) *The Principia: Mathematical Principles of Natural Philosophy* (1687).

The work of Albert Einstein in the early part of the twentieth century has shown that there is a connection between time and space that is perhaps not immediately apparent in everyday life. The complexities of General and Special Relativity are beyond the scope of this volume. Isaac Newton in his quotation from his *Principia* (see above) provides us with a usable definition for the study of geological time. The key word is duration – for the purposes of this study we can consider time as a succession of events, placed in the order of their happening, with regard to their duration in time.

By the late nineteenth century the stratigraphical column was more or less complete and in a form we would recognize today. Geologists, mining engineers and surveyors made correlations and identified patterns in the sediments and fossils. They realized that a number of major geological boundaries were recurrent and applicable across the world. After centuries of fitting geological time into an unyielding Creationist framework supplied by the various church authorities, the idea of the immensity of geological time was now close to scientific orthodoxy.

One of the principal beneficiaries of this changed climate of opinion was Charles Darwin, who was then able to use the greater time period available to develop his ideas on evolution and the origin of species. The great advances in stratigraphy in the nineteenth century meant that the beds were now largely placed in the correct order according to their *relative age*. The next great challenge was to work out the duration of each of the periods in years – their *absolute age*. As progress was made establishing the timescale of the Phanerozoic or visible life, it was becoming obvious that there was a considerable time period involved in the supposedly fossil-free rocks of the Precambrian – or the *Pre-Cambrian* – meaning before the Cambrian, as it was originally designated. Along with dating the Phanerozoic or visible life eon, unravelling the extent and sub-divisions of the Precambrian and gauging the age of the Earth were the next major tasks for a new generation of geologists to undertake.

Determining the age of the Earth

Speculation on the age of the Earth, often contextualized as the date of creation, has been a subject of enquiring minds for thousands of years, exercising the philosophers of all the early civilizations. In European society the speculation intensified during the Scientific Revolution beginning in the sixteenth century, and which we discussed in some detail in chapter 2. During this period advances in scientific understanding, particularly in the fields of mathematics and astronomy, led to challenges to the religious orthodoxy of the time, which dictated a model for the origin of the Earth based on the account given in the Old Testament. It could be said, in fact, that the basis of the scientific revolution was recognition of how little was actually known about the world, thus providing an incentive for scientific progress.

The response of the church authorities to this perceived attack on their tenets of faith, as well as acts such as the enforced recantation by Galileo, was to publish a large number of calculations of the age of the Earth based on the text of the Old Testament. The most famous of these was by James Ussher (1580–1656), Archbishop of Armagh in Ireland.

Ussher is famous for calculating the date of the Creation as the night preceding the 23rd day of October, 4004 BCE. His *Annales Veteris Testamenti* (Annals of the Old Testament) was published in 1650 and became the definitive biblical chronology for the period. The first page of the Annales

40

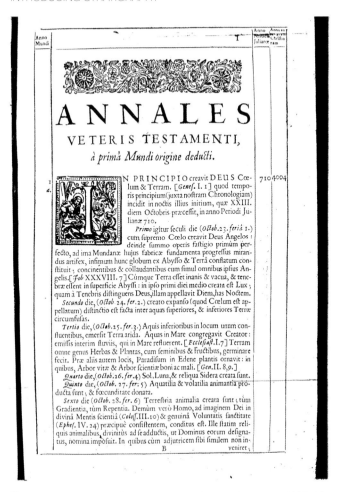

Figure 4.1 James Ussher's *Annales Veteris Testamenti* (1650).

Veteris Testamenti includes the date 4004 BCE in the margin (see Fig. 4.1). The theories of Earth's origin that appeared in the seventeenth century were not concerned solely with the age of the Earth, but also with explaining the observed features of the world in terms of the story of Creation and Noah's Flood as detailed in Genesis. To some extent there is still an on-going debate today between those like Descartes, who attempted to explain the Earth in terms of its observed characteristics, and those who still resist the idea of an old Earth and advocate an origin based on the Old Testament narrative.

Attempts to determine the age of the Earth: accumulation processes

Chapter 1 referred to the Greek historian Herodotus and his comments on the link between time and the accumulation of sediments in the Nile Delta in Egypt. This probably represents the first recorded attempt at using accumulation processes to estimate age, and was the precursor to many such attempts to do so using a range of methods. The thinking behind these methods is the same in each case. Since the basis of Uniformitarianism is that processes operating today are essentially the same processes that have always applied, then the rates of, for example, sediment accumulation could be used to calculate the age of the Earth if the rate of deposition and the total thickness accumulated were known. This method was attempted by John Phillips in the 1860s.

Using rates of deposition and erosion for the Ganges Delta in the Bay of Bengal, Phillips calculated a total thickness of the fossiliferous sediments of the stratigraphic column of 72,000 feet (about 22,000m) which he estimated gave an age of the Earth of approximately 96myr. Like all such attempts using sediment accumulation, it failed to take into account factors such as variations in accumulation rates for different sedimentary environments and the significance of breaks in the sedimentary record. Although we would now consider 96myr as a gross underestimation of the age of the Earth, it nevertheless was an important contribution to the debate on the extent of geological time. The age of the Earth was becoming a major cause of controversy in scientific circles at that time in the mid-nineteenth century.

Another imaginative approach to calculating the age of the Earth was originally proposed by the English astronomer Edmund Halley (1656–1742), whose name is best known for its association with the periodically returning comet. In 1715 he suggested estimating the age of the Earth by measuring the rate of salt accumulation in the world's oceans and working out the age from the current ocean salinity. Halley reasoned the current salinity of the oceans could not have developed in the comparatively short period required to fit with Archbishop Ussher's biblical timetable, but that, on the other hand, it could not have been a very long period or the oceans would be as saline as the Dead Sea. As it happened, Halley never completed his salinity project and it was left to an Irish scientist, John Joly (1857–1933), of Trinity College Dublin, to revisit the idea in 1899. Joly made the assumption that the oceans were a closed system with no salt being lost once it had accumulated. He calculated that the oceans developed around 80–90 million years ago. This was also an underestimate, since we now know the ocean is not a closed system, and there is a continual loss of salt to various sedimentary and meteorological processes. For example, during the course of plate-tectonic events such as the subduction of the sea bed into the mantle at destructive plate margins, salt will be removed from the oceans. More mundanely, a short spell parked near the sea during windy weather will result in the car windscreen being covered with a layer of salt from spray blown off the surface of the sea.

The cooling Earth

In the late seventeenth century Archbishop Ussher's age for the Earth of around 6000 years was the orthodox view of the matter. However, as with the earlier phase of the Scientific Revolution in the previous century, changes were in the offing, with a new generation of scientists prepared to challenge the authority of the churches and to put forward the case for rationalism. One such was Georges-Louis Leclerc, Compte de Buffon (1707–88).

Buffon was a French naturalist and mathematician who investigated a wide range of topics in the natural world. He believed the Earth to be much older than the figure put forward by Archbishop Ussher. In the late 1770s he undertook a series of experiments, in which he heated a number of iron balls to near melting point and measured the time it took for them to cool down. By extrapolation he arrived at a figure of nearly 100 thousand years (kyr) for the temperature of the balls to fall to the Earth's present temperature. By adjusting the composition of the iron spheres to more closely resemble the composition of the Earth, he later modified this age to 75kyr. Buffon's ideas attracted much opprobrium from the religious establishment of the time and his work seemed to make little impact. However, as with Halley's work on salt accumulation some 50 years earlier, his ideas were to be revived many years later. In Buffon's case the scientist who resurrected the idea of a cooling Earth was probably the most eminent and influential scientist of the late-Victorian age in Britain, William Thomson, Baron Kelvin of Largs (1824–1907). Lord Kelvin's influence on the development of physics during the second half of the nineteenth century was immense and he was a major contributor to the formulation of the Second Law of Thermodynamics.

We have already considered the Second Law of Thermodynamics in a geological context – the concept of Time's Arrow and the one-way direction of increasing entropy or disorder being used to separate past and future time (see chapter 3). In 1862 Kelvin attempted to calculate the age of the Earth by applying thermodynamic principles to heat flow, and thus follow on from the Compte de Buffon's work on heated iron spheres some 80 years earlier.

The basis of his calculation was that the Earth and the Sun were originally equally hot and both had been cooling ever since, and also that the Earth was incapable of generating new heat. Initially he calculated an age of 98myr but by 1899 he had revised the figure down to 20–40myr, with the probability that the age was in the lower part of this range. From the start his views on the likely age of the Earth were controversial among the scientific community, particularly the geologists. In 1869 Thomas Henry Huxley, then president of the Geological Society of London, questioned his proposed timescale of less than 100myr. This dissent from geologists and then biologists continued for most of the rest of the nineteenth century. However, such was his reputation that his influence on the controversy remained considerable until the next major discovery in physics totally changed the terms of the argument. A man of trenchant views, he is famously quoted as saying *In science there is only physics, all the rest is stamp collecting*, which probably did little to endear him to scientists working in other fields. However, he also went on record to say that heavier-than-air flying machines are impossible, which along with his views on the age of the earth, meant that he was on the wrong side of at least two major developments in science.

The discovery of radiation

Throughout the controversy generated by Kelvin's proposals for the age of the Earth based on cooling rates, a major assumption had been that no new heat was involved in the cooling history of the Earth. As a result of the research by a husband and wife team, Pierre and Marie Curie, working in France, it became known that there is more heat in the Earth than Kelvin was aware of. Kelvin's model of the cooling Earth was therefore fundamentally flawed.

In 1895 Wilhelm Conrad Röntgen, a German physicist working with an early type of cathode ray tube (found in older types of television sets), discovered that a previously unknown form of energy came from this device when an electrical current passed through it. He realized it was capable of passing through opaque materials and he named the energy source *X-rays*.

Shortly after the discovery of X-rays, Pierre and Marie Curie investigated pitchblende, an impure form of the mineral uraninite, the commonest ore of uranium. They were following on from work by another French physicist, Henri Becquerel (1852–1908) who had discovered in 1896 that uranium compounds spontaneously emitted energy with similar powers of penetration to Röntgen's X-rays. In the course of their research they discovered two new elements, radium and polonium, which also spontaneously emitted energy. Marie Curie coined the term 'radioactivity' for this phenomenon. The significance of these new energy sources in the context of the age of the Earth was of course that energy equals heat, and this immediately negated Kelvin's assumption that the Earth had no further heat source. In fact by 1904 another of the pioneers of radiation physics, Ernest Rutherford (1871–1937), stated that the amount of heat emitted from radium is sufficient to melt more than its weight of ice per hour. Over the next few years Rutherford developed a general theory of radiation, the **disintegration theory**, which stated that the atoms of radioactive elements are unstable and constantly disintegrating to form another element with the emission of energy. The new element may decay further until a stable element is finally formed. In his 1904 book *Radioactivity*, Rutherford provided evidence that the rate of radioactive decay is constant and not affected by external conditions. He further suggested the end product of this decay was the gas helium, and that if estimates could be made of the amount of helium trapped in the minerals of the rock, then estimates of the age of those minerals could be made. This eventually was to be the basis of a method for the quantification of geological time.

The result of these discoveries at the end of the 1890s and the beginning of the twentieth century was to a large extent to free geologists from the constraints imposed by Kelvin's cooling Earth model. Here, after all, was an undoubted source of internal heat in the Earth which allowed much longer periods to cater for the concept of evolutionary changes and species development, and was to be welcomed. In 1905, Robert John Strutt, later to be Baron Rayleigh, following on from Rutherford's speculation about helium, analysed the helium content of a radium-bearing rock and estimated its age to be two billion years (byr). The scale of this figure was confirmed by work in the USA by an American chemist, Bertram Borden Boltwood, who using lead as a likely decay product, calculated the age of a number of minerals and found ages ranging from 400myr to 2.2byr. The sheer scale of these ages was proving to be a problem to many geologists. They were having difficulty adjusting from having not enough time for geological processes to having ages that were orders of magnitude greater than anything previously considered. The stage was set for the first contribution to geological science from a man whose influence on geological thought in the twentieth century was both profound and lasting.

Arthur Holmes and radiometric dating

Arthur Holmes was born at the beginning of the last decade of the nineteenth century, and died in 1965 just as the so-called revolution in the earth sciences, Plate Tectonics, was gaining momentum. Few geologists can have made such an immense personal contribution to the development of this new radical view of geology than Holmes, but here we are concerned with his pioneering work on dating rocks using radiometric methods.

At the beginning of his career, Holmes worked in Strutt's laboratory in Imperial College on the problems of applying radiometric techniques to dating. In 1911, at the early age of 21, Holmes published his first scientific paper in the prestigious *Proceedings of the Royal Society*. Using the uranium–lead decay process he produced an age of 370myr for a Devonian rock. This, along with his book entitled *The Age of the Earth*, which he published to widespread acclaim in 1913, consolidated his reputation as the leading expert in the field. He had embarked on a major aspect of his life's work, to formulate an accurate timescale for the geological column. For the first time absolute age dates were being proposed for rocks within their stratigraphical context.

In 1913 Frederick Soddy, a research collaborator of Ernest Rutherford and co-author of the disintegration theory of radioactive decay that had initiated estimates of the age of minerals using radiometric methods, discovered **isotopes**.

All elements contain protons and neutrons in the atomic nucleus and electrons that orbit around the nucleus. In each element the number of protons is constant while the number of electrons and neutrons can vary. Atoms of the same element but with different numbers are called *isotopes* of that element. The element hydrogen exists in three isotopic forms:

- Hydrogen-1, mass number 1
- Hydrogen-2, also known as deuterium, mass number 2
- Hydrogen-3, also known as tritium, mass number 3
  See Figure 4.2.

The mass number is the number of protons plus neutrons. Hydrogen therefore has three isotopes with mass numbers 1, 2 and 3. A nucleus of carbon has six protons but can have six, seven or eight neutrons. Carbon is found therefore in three isotopic forms, carbon 12, carbon 13 and carbon 14. By convention these are shown as ^{12}C, ^{13}C and ^{14}C where the superscript number corresponds to the mass number of the isotope.

Holmes was unaware of the existence of isotopic forms of uranium and lead on which he had based his 1911 age date, and so this figure could not have been entirely accurate. In these pioneering days of radiometric dating there were a number of uncertainties concerning the extent of isotopic variations among the elements and about the specifics of many of the decay series under investigation. Resolution of many of these problems had to wait until the development of better

analytical equipment and techniques, notably the mass spectrometer, which became important from the mid-1920s onwards. This equipment allows the identification of isotopic form of elements and also their relative abundance.

Radiometric dating: principles and methodology

Atoms of elements that display radioactivity are inherently unstable and break down or decay to form a new element. This is conventionally described as **parent atoms** decaying to form **daughter atoms**, resulting in a **decay series**. The longer the process has been going on for, the greater will be the number of daughter atoms and conversely the smaller the number of parent atoms. Since the rate of decay has been shown to be constant, the ratio of parent to daughter atoms is a measure of how long the process has been going on and therefore of the age of the mineral.

An element commonly used to calculate age dates is uranium. Two of the isotopic forms of uranium have mass numbers 235 and 238 – ^{235}U and ^{238}U. An atom of ^{235}U consists of 92 protons + 143 neutrons, while ^{238}U has 92 protons + 146 neutrons.

The **half-life** of a radioactive element is the time taken for half the initial number of parent atoms to decay to daughter atoms. For example, the isotope of uranium ^{235}U which decays to give the form of lead ^{207}Pb has a half-life period of 704myr. Other examples are ^{40}K (potassium) which decays to ^{40}Ar (argon) with a half-life of 1193myr, and ^{87}Rb (rubidium) decays to ^{87}Sr (strontium) with a half-life of 48,800myr.

Table 4.1 below shows decay series commonly used to calculate radiometric dates.

The relationship between the number of parent and daughter atoms with time is shown in Figure 4.3. Starting with a hypothetical number of 100 parent atoms, after one half-life the number of parent and daughter atoms are equal, the ratio is 50:50. After a further half-life period the parent to

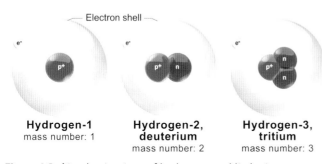

Figure 4.2 Atomic structure of hydrogen and its isotopes.

Hydrogen-1
mass number: 1

Hydrogen-2, deuterium
mass number: 2

Hydrogen-3, tritium
mass number: 3

Electron shell

Table 4.1 Radioactive isotope decay series used to calculate radiometric dates.

Parent isotope	Daughter isotope	Half-life period	Dating range	Occurs in
Uranium 238 (^{238}U)	Lead 206 (^{206}Pb)	4500myr	10–4 600myr	Zircon
Uranium 235 (^{235}U)	Lead 207 (^{207}Pb)	704myr	10–4 600myr	Uraninite
Potassium 40 (^{40}K)	Argon 40 (^{40}Ar)	1193myr	0.2-4 600myr	Micas, hornblende, slates, some sediments
Rubidium 87 (^{87}Rb)	Strontium 87 (^{87}Sr)	48800myr	10–4600myr	Micas, feldspars, granites, gneisses

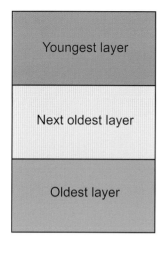

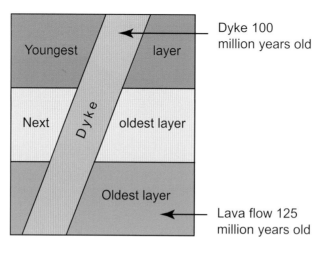

Figure 4.5 Law of cross-cutting relationships.

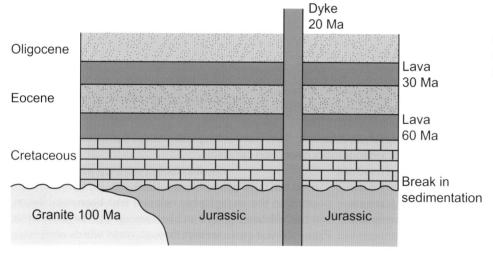

Figure 4.6 Calibrating the stratigraphic column using radiometric dates.

possible to determine its maximum age. The Cretaceous, which overlies the Jurassic, has been deposited on top of the Jurassic after a break in sedimentation. It must therefore be *younger* than 100myr, but is overlain by a lava flow which is dated at 60myr, and this gives an upper age for the Cretaceous.

Immediately above the 60myr lava are the sediments of the Eocene and the Oligocene, separated by another lava flow dated at 30myr. This sequence means that the Eocene sediments are less than 60myr old, but older than 30myr. The Oligocene is less than 30myr old because it lies above the 30myr lava flow, but along with the rest of the sequence it is cut by a 20myr old dyke, so the whole sequence is older than 20myr.

To summarize:
- the whole sequence is older than 20myr;
- the Oligocene is older than 20 but younger than 30myr;

- the Eocene is older than 30 but younger than 60myr;
- the Cretaceous is older than 60 but younger than100myr;
- the Jurassic is older than 100myr, but its maximum age cannot be determined from the available evidence.

It should be noted that although the lava flows are not sediments, for the purposes of the Law of Superposition they are regarded as part of this horizontal succession of rocks because they are conformable members of the sequence.

In the years since the mid-twentieth century the calculation of radiometric age dates has become more precise and accurate, and by using thousands of age dates applied to thousands of successions all over the world, a geological timescale has been produced that provides a calibrated structure against which every geological event can be referenced.

Calculating a radiometric age for the Earth and deconstructing the Precambrian

The age of the Earth is 4.54byr old. This is based on age dates obtained from meteorites, which are believed to represent the primitive material from which the solar nebula was formed. Because the Earth separated into the core, mantle and crust at an early stage and was then subjected to plate-tectonic processes, rock samples from Earth are not considered reliable sources for a direct date for the formation of the Earth. The oldest rocks on Earth are to be found in the cratons, the old, geologically stable cores of the continental areas: for example, the Pilbara Craton of West Australia, the Canadian Shield Craton or the Congo Craton in Africa. All of these areas are characteristically composed of crystalline metamorphic rocks, often referred to as **basement rocks**. These areas contain some of the oldest rocks on Earth, dated at more than four billion years, which implies that, at a very early stage in the Earth's evolution, there was a stable crust. The ability to gather data at this level of detail means it has been possible to deconstruct the previously undivided Precambrian and gain some insight into a complex geological story involving ocean formation and complex continent movements culminating in the building of supercontinents. The data has also helped in the understanding of how life originated and evolved over this vast time period.

The term Precambrian is used to denote the oldest rocks on the planet and the designation was for those rocks which were older than rocks of Cambrian age. By the end of the nineteenth century, Cambrian rocks were considered to be the oldest fossil-bearing sediments on Earth. One of the first consequences of the extent of geological time revealed by radiometric dating was the realization of the sheer scale of Precambrian time. It was clear that a substantial period of Earth history had occurred before the arrival of the animals with shells that were the diagnostic feature of the Cambrian period. The Precambrian accounts for almost 90% of the total history of the Earth. See Figure 4.7.

Not only has radiometric dating shown the extent of the Precambrian, but since the beginning of the twentieth century research has revealed a diverse fauna in the Precambrian. This

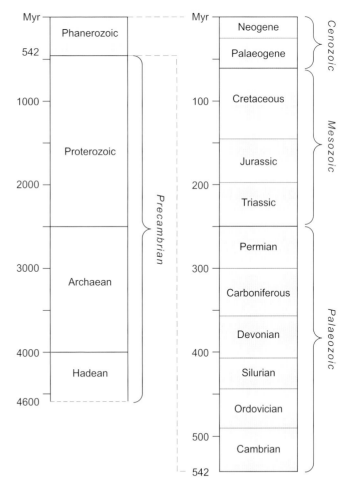

Figure 4.7 The stratigraphical column: the left-hand section shows the true-scale extent of Precambrian time.

Why the stratigraphic column is subdivided

We have established the stratigraphic column on a world-wide basis, with its by now familiar subdivisions (see Fig. 4.9). These range from the oldest rocks in the Precambrian, including the most recently designated period, the Ediacaran, to the Quaternary. It is an interesting time to be studying stratigraphy. The most recent epoch of the Quaternary, the Holocene, is probably about to be succeeded by a new epoch designated the **Anthropocene**. Named after the Greek word *anthropos*, meaning humans in general, the new term recognizes the changes in the Earth's environment made by *Homo sapiens*. At this stage of the book it would be useful to examine the complete timeline of Earth history. What changes would have been visible on the Earth's surface as time progressed from the earliest Precambrian to the present day and the proposed new epoch the Anthropocene?

The major sub-divisions of the stratigraphic column are defined by important events, many of which would result in changes clearly visible from space. Phenomena such as catastrophic meteorite strikes would have an immediate and visible impact on the Earth. Other changes such as the distribution of continents on the surface of the globe play out over hundreds of millions of years producing the Supercontinent Cycle. This is the process whereby the continents on the Earth's surface amalgamate periodically into a single landmass, before splitting and separating again (see chapter 3). Modern rates of sea-floor spreading are of the order of a few centimetres per year, but it is likely that in the Precambrian spreading rates may have been two to three times greater than at present. These faster rates were the result of a higher temperature mantle caused by greater heat production in the mantle from faster rates of decay of radioactive isotopes in the early days of the Earth. Plate tectonics provides us with a structural framework for the planet. Following its early differentiation into a central core surrounded by the mantle with an outer skin of continental or oceanic crust, the complex choreography of continent formation and rifting and ocean opening and closing began, driven by convection cells within the mantle. (See Figures 3.1 to 3.4). Seen from space, the Earth would thus present a constantly changing appearance as continents moved across its surface.

Global sea-level rises and falls brought about by these changes in plate-tectonic processes or the incidence of glaciations make for significant changes in sediment production and distribution. Indeed, it is variation in sea level – the transition from sea to land and back again – that generates the sedimentary sequences that make up much of the geological record and record the passage of geological time. In addition, the periodic occurrence of ice ages from Precambrian times onward has modified landforms and also had profound implications for the development and evolution of life on Earth.

Other changes through time that have had sufficient impact to warrant sub-divisions of the stratigraphic column are the origin and evolution of life, along with the mass extinction events that periodically occurred. These accounted for huge losses of plant and animal groups, both in the oceans and on land. The Proterozoic Eon of the Precambrian is defined as the eon of *early life*, while the succeeding Phanerozoic Eon is the eon of *visible life*, now generally taken as involving complex life forms.

The aim of this chapter is to review the stratigraphic column in the light of phenomena such as these and to construct a history of the Earth as evidenced by the main sub-divisions of the rock succession. We are looking at events and processes over a 4.5byr period that formed the geological succession applicable today all over the world. In considering such an exercise it is important to emphasize the length of geological time involved. The stratigraphic column depicted in Figure 4.9 showed that the Precambrian extended from 4.5bya to 542mya, representing almost 90% of the total age of the Earth. Although the Precambrian represents such a large percentage of geological time, rocks of that age tend not to be well exposed on the Earth's surface. This is because they may have been eroded and recycled as part of the plate-tectonic

Figure 5.4 Boulder clay cliff.
©Graeme Eyre/Shutterstock

Figure 5.5 Granite clast in the Port Askaig tillite, County Donegal, Ireland.

Evidence for this glaciation includes extensive tillite deposits and rocks with glacially derived **striations**, scratches left on rock surfaces as the ice moved across the surface. See Figure 5.7 showing the Nooitgedacht Glacial Pavement in South Africa, where the Karoo Ice Age left a glacially scoured, striated bedrock platform, which was then utilized by early (late Holocene) inhabitants of the area as panels for rock engravings or petroglyphs. The geographical extent of the glaciations is shown by similar features from the Paraná Basin, Brazil.

Evidence for late Precambrian glaciations in the rocks of Britain and Ireland is shown by the Port Askaig Tillite. Port Askaig is on the east coast of the island of Islay on the west coast of Scotland. The tillite can be found from western Ireland to the northeast of Scotland.

Further glaciations in the Proterozoic occurred between 850 and 630bya and there is evidence that the emergence of more complex life forms may have removed carbon dioxide from the atmosphere and thus triggered a severe glacial event with further episodes of Snowball Earth. The end of this prolonged glacial period heralded the Ediacaran Period (635–541myr) which marks the transition between the Proterozoic and the Phanerozoic. This cyclic variation in global temperature was

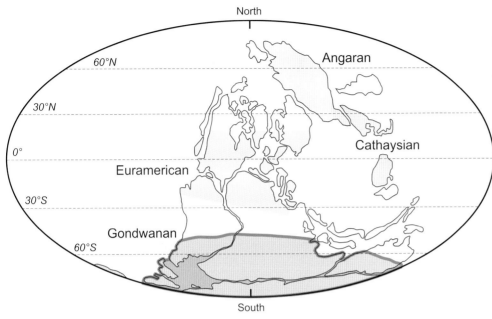

Figure 5.6 Approximate extent of the Karoo glaciation (in blue) over Gondwana during the Carboniferous and Permian.

Figure 5.7 Te Nooitgedacht Glacial Pavement in South Africa showing a glacially scoured and striated surface with petroglyphs or rock engravings.

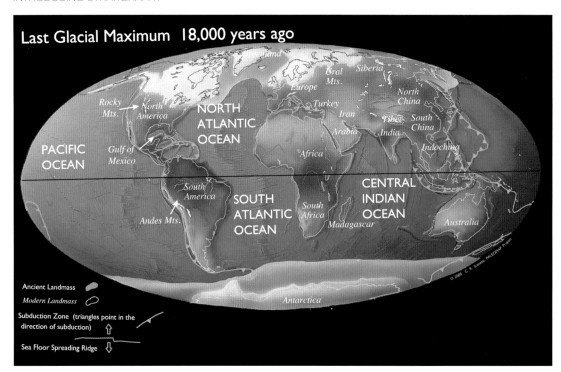

Last Glacial Maximum 18,000 years ago

Ancient Landmass

Modern Landmass

Subduction Zone (triangles point in the direction of subduction)

Sea Floor Spreading Ridge

Figure 5.8 Northern Hemisphere glaciations during the last Glacial Maximum about 18kya.

to continue right through the Phanerozoic up to the present day, as shown by Figures 5.2 and 5.3.

Figure 5.8 shows the extent of the recent, Pleistocene, glaciation in the northern hemisphere, viewed from directly above the North Pole.

Sea-level changes

Charles Lyell used this image of the 2000-year-old Macellum or market building of the city of Pozzuoli in southern Italy (also known as the Temple of Serapis) as the frontispiece of his *Principles of Geology*, published in 1830. The three columns in the centre field of the image show bands of borings by *Lithophaga* (literally rock-eating) bivalve molluscs. The position of these bands showed how the site had gradually subsided below sea level and then re-emerged. To Lyell this was proof that ongoing geological processes can have significant effects over long periods of time, in agreement with his concept of Uniformitarianism (see chapter 3).

Many of the changes and variations seen in the stratigraphic column are the result of rising or falling sea levels.

The frontispiece of Lyell's *Principles of Geology*.

The role of these changes in the processes of sedimentation is profound, and a later chapter will deal with these in detail. Change in sea level results from alteration in the volume of sea water or variations in the volume of the basins filled by sea water. The thermal expansion or contraction of sea water will cause a change in volume, as will ice cap growth and melting. Change in the volume of basins is due to tectonic processes.

Sea level, or *relative* or *mean* sea level, as it is also referred to, is the average height of the ocean's surface between high and low tide, measured against a local datum. In the UK, for example, height above sea level is measured from the *Ordnance Datum*, considered to be the mean sea level as shown on the marker at Newlyn Pier, Cornwall. A rise in global sea levels is described as **eustatic**, measured between the surface of the sea and a fixed datum, usually the centre of the Earth. It is important to recognize that local changes may be quite different.

Consider for the moment sea level changes due to glaciations. The Earth is currently in an interglacial period. Over the period covered by this ice age, global sea level was, on average, around 130m lower than it is today due to the enormous amount of sea water stored in ice caps and glaciers. There are also important differences in the rates at which sea level changes occur. Changes due to the formation or melting of ice caps happen relatively quickly, occurring over periods of perhaps tens or hundreds of thousands of years. The vast weight of the ice, up to several kilometres thick in places, causes the Earth's crust to downwarp by deforming the underlying mantle due to a process known as isostatic readjustment (see Fig. 5.9). When the ice has melted, the crust is unloaded

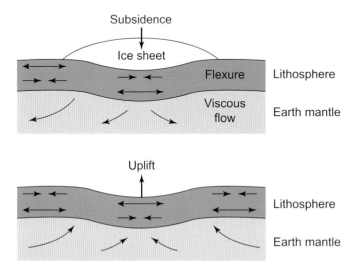

Figure 5.9 Isostatic readjustment of the crust after melting of ice sheet causing a fall in sea level.

and the underlying mantle flows back in and raises the crust, and the sea level falls. This is known as an isostatic sea-level change and is a relative change since the land is rising, but the amount of water is not increasing. A rise in sea-water temperature will also contribute to a rise in sea level, since water expands at higher temperatures.

The post-glacial rebound effects on the land level of Britain and Ireland are shown in Figure 5.10. Regions shown in green are still rising; those shown in brown are sinking, while those areas coloured orange are stable or near stable. In contrast to the relatively rapid sea-level changes due to isostatic adjustments, those changes associated with the movement of tectonic plates which alter the configuration of ocean basins can take tens or hundreds of millions of years. For example, the collision of the Indian and Eurasian plates that formed the Himalayas (see Figs 3.6 and 3.7) when a large volume of continental crust was subducted or forced under Tibet; the size of the ocean basin was greatly increased, causing a lowering of global sea level.

As a general rule, sea level is lower when the continents are close together and higher when they are separated. This is clearly illustrated by Figure 5.11. The lowest sea levels are seen between the late Carboniferous and the Jurassic, when the continents formed the supercontinent Pangaea (C,P Tr,J). Global sea level then rose steadily through the Cretaceous (K) as the supercontinent fragmented and the Atlantic opened. This process involved active, young, hot mid-ocean ridges, which occupy a greater volume than older non-active ridges that have cooled and shrunk.

The most up-to-date chronology of sea-level change through the Phanerozoic shows the following long-term trends:
- Gradually rising sea level through the Cambrian consistent with rifting associated with the opening of the Iapetus Ocean;
- Global sea-level drop at the end of the Cambrian attributed to the start of subduction within the Iapetus Ocean;
- Relatively stable sea level in the Ordovician, with a large drop associated with the end-Ordovician glaciations;
- Relative stability at the lower level during the Silurian with the change towards the end of the Silurian coincident with the start of subduction in the Rheic Ocean, which had opened to the south of the Iapetus Ocean and was now beginning to close;

57

Figure 5.10 Map of the post-glacial rebound effects on the land level of Britain and Ireland.

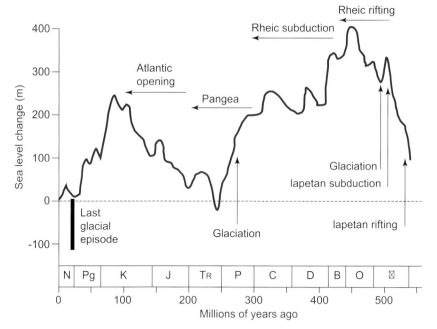

Figure 5.11 Global sea level curve for the Phanerozoic (after Hallam, 1992).

- A gradual fall through the Devonian, continuing through the Carboniferous to the end-Carboniferous glaciation;
- A gradual rise until the start of the Permian, followed by a gentle decrease lasting until the end of the Triassic, coinciding with the existence of the supercontinent Pangaea;
- The rise in sea level in the Jurassic and Cretaceous coincides with the opening of the Atlantic and associated rifting.

The role of sea-level changes in the branch of stratigraphy known as sequence stratigraphy will be discussed in detail later in the book.

Meteorite impact

The early history of the Earth involved a phase of heavy bombardment by asteroids and meteorites as the growing planet attracted Solar System debris under the influence of gravity. To see the results of such a bombardment, one only needs to examine the surface of our nearest neighbour, the Moon (see Fig. 5.12).

Chapter 3 discussed the roles of Uniformitarianism and Catastrophism in geological history, and the meteor impact at Chicxulub in Mexico was used as an example of a catastrophic

Figure 5.12 Impact craters on the moon.

that a significant extinction event took place around 542mya. This may have been caused by changes in oceanic circulation leading to anoxic conditions, or may have been due to the overlap in time with the metazoan animals that were soon to dominate all environments. The so-called Cambrian explosion was probably due to the radiation of animal species filling environmental niches left vacant by the extinction of so many species at the end of the Ediacaran Period.

The Ediacaran fauna represent the oldest multicellular organisms found in the stratigraphic column, but as discussed earlier, the oldest known fossils are stromatolites, produced by the activity of water-living cyanobacteria that were capable of photosynthesis (see Figs 4.10 and 4.11). By their precipitation of calcium carbonate grains, stromatolites not only began the process of oxygenation in the early atmosphere and built up the ozone layer, but were important ecologically as the first reefs. Their activity eventually led to sufficient oxygen accumulating in the atmosphere to allow the development of more complex life forms. This gradual increase in the level of oxygen in the early atmosphere and the formation of the ozone layer had important implications for the development of life on Earth, particularly with respect to the colonization of land areas by increasingly complex plants and animals, leading eventually to the group we belong to, the mammals.

Figure 5.16 shows the concentration of oxygen in the atmosphere for the last 1000myr. The red line represents the present concentration of 21%. By the beginning of the Cambrian (around 500mya) oxygen levels had jumped from the 3% levels of the late Proterozoic to around 12%. This increase corresponds with the Cambrian explosion of life forms referred to earlier in the chapter. The spread of plants over the land during Silurian and Devonian times (around 450–400mya) enabled a further rapid increase in the atmospheric oxygen content. Carbon burial in the Carboniferous (around 350–300mya) allowed oxygen levels to rise to 35%. The significant drop in atmospheric oxygen level that coincided roughly with the Permian–Triassic boundary was caused by extensive and prolonged volcanic activity. After the low of the Permian–Triassic, levels of oxygen slowly recovered, reaching 30% in the Cretaceous (around 100mya) before declining to the present value of 21% by 25mya. Oxygen levels have stayed around this figure since then, with only minor fluctuations.

Life has evolved into three groups of closely related organisms – Archaea, Bacteria and Eukaryota. The first two are relatively simple cells surrounded by a membrane and cell wall, with a circular strand of DNA containing their genes, and are called **prokaryotes**. The stromatolites discussed earlier in the chapter are prokaryotes, and were well developed by around three billion years ago. **Eukaryotic** cells are more complex than prokaryotes and contain a visible nucleus containing the DNA and cell components such as mitochondria, which act as miniature power plants (see Fig. 5.17).

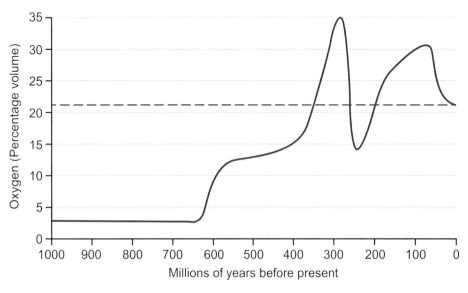

Figure 5.16 Atmospheric oxygen concentration over the last 1000myr.

Typical prokaryote cell

Typical eukaryote cell

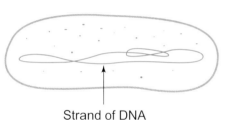
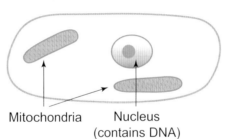

Strand of DNA

Mitochondria   Nucleus
(contains DNA)

Figure 5.17 Components of typical prokaryote and eukaryote cells.

Virtually all life forms today are eukaryote. As the level of oxygen in the atmosphere increased it became possible for organisms to use oxygen directly in their respiration, and there is evidence to suggest that eukaryotic cells similar to those found in higher plants and animals today evolved soon after this period. By the time of the widespread appearance of eukaryotes at 1.8 billion years ago, oxygen concentration had risen to 10% of present atmospheric level (PAL). The complex eukaryotic cell ushered in a whole new era for life on Earth, because these cells evolved into multicellular organisms. A second major peak was reached by 600mya which raised atmospheric oxygen levels to 50% PAL. It was denoted by the first appearance of skeletons. The Ediacaran fauna from around this time shows complex life forms, diverse soft-bodied animals, with hard-shelled creatures appearing towards the beginning of the Cambrian. The development of skeletons and hard shells meant that there was an increased chance of organisms being preserved after death as fossils. An animal consisting of only soft parts will most likely disappear soon after death without leaving a trace. This evolutionary advance in animal physiology means that while the fossil record is always incomplete, overall fossils are very rare occurrences; fossil evidence becomes more complete after the Cambrian. See Figure 5.18.

Mass extinction events

Land areas seem to have been devoid of plants and animals until cyanobacteria and other microbes formed prokaryotic mats that covered terrestrial areas. The arrival of more complex plant and animal life on land was still some millions of years in the future. The Cambrian Explosion of life marks the start of the Palaeozoic (*old life*) Era.

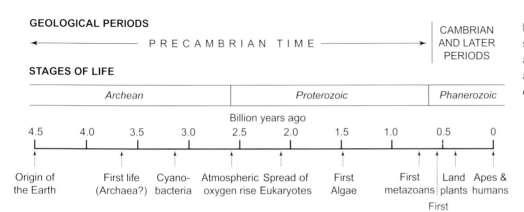

Figure 5.18 Time chart showing Precambrian time as a proportion of Earth history and major stages in the evolution of life.

pressures is called **maturation** and will be discussed further in later chapters. Since the time taken to convert dead plankton to hydrocarbons is of the order of tens of millions of years, it is obvious that oil and gas are classified as non-renewable resources.

Just as the beginning of the Jurassic had been marked by a mass extinction, so was its end. The event did not match in magnitude the Triassic–Jurassic catastrophe, nor is it considered among the most important mass extinction events of the Phanerozoic Eon (shown in Fig. 5.19). Nevertheless many dinosaur species, including the enormous sauropods such as *Diplodocus* died out, as did many forms of ammonites, marine reptiles and more than 80 percent of marine bivalve species. The cause of this extinction is unknown.

The Cretaceous Period (145–65mya) is the last of the three periods of the Mesozoic. Throughout the Jurassic, global sea levels had been steadily rising, mainly as a result of increased rifting of the supercontinent Pangaea. Increased plate-tectonic activity in the Cretaceous with the formation of the North and South Atlantic oceans meant that mid-ocean ridges

expanded greatly and thus displaced the water in the ocean basins, leading to higher global sea levels. This trend continued in the Cretaceous and the world's oceans were at a higher level during most of the Cretaceous than at any other time in Earth history. It is estimated that sea level was around 100–200m higher than today in the early part of the Cretaceous, rising to around 250m higher by the end of the period. The volcanism associated with this sea-floor spreading probably accounts for the warmer and more humid climate that prevailed in the Cretaceous.

The term 'Cretaceous' is derived from the Latin *creta* for chalk. Chalk is a fine-grained form of limestone, formed from the calcareous shells of **coccoliths**, which were floating algae common in Cretaceous seas. The high eustatic sea levels, coupled with the warm climate, meant that a high percentage of the world's continents were inundated by warm, shallow seas resulting in extensive and thick deposits of chalk in many parts of the world. At maximum sea level there was probably less than 20 percent of the Earth's surface covered by land, compared with around 30 percent today (see Fig. 5.29).

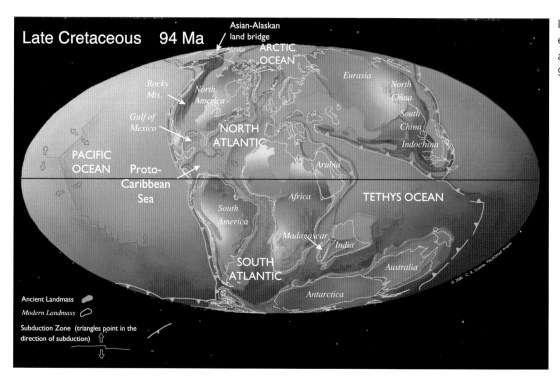

Figure 5.29 Shows the extent of shallow seas across the globe 94mya.

During the Cretaceous an important development was the first appearance of flowering plants in the middle Cretaceous. Around the same time insects were beginning to diversify, including groups such as bees, wasps, grasshoppers and ants.

The end of the Cretaceous is delineated by the Cretaceous–Palaeogene boundary, a marker horizon associated with a mass extinction that lies between the end of the Mesozoic Era and the succeeding Cenozoic Era, and was discussed earlier in the chapter. This marker horizon, which occurs worldwide, is a layer of clay, rich in the element iridium (see Fig. 5.30). As mentioned earlier, iridium is extremely rare in the Earth's crust, but is much more common in meteorites. It is thought, therefore, that this layer is the result of the impact of a major meteorite with the Earth at Chicxulub in the Yucatan Peninsula, Mexico (discussed in chapter 3 – see Fig. 3.10). This meteorite is estimated to have been at least 10km in diameter and produced an impact crater more than 180km across.

It is widely held that this impact accounted for the demise of the dinosaurs, along with probably half of all species on Earth at the time. Whether or not this was the sole cause of the mass extinction or merely one of a range of factors, the effects of such a collision would have been catastrophic. The shock waves generated would have caused earthquakes and volcanic eruptions triggering huge tsunamis across the world's

oceans. The atmosphere would have been full of super-heated dust and ash leading to long-term effects on plant and animal life. As with the end-Permian mass extinction, which coincided with the volcanic eruptions of the Siberian Trap basalts, the end-Cretaceous event also overlaps with a major volcanic episode. The Deccan Trap basalts of northwest India poured out around one million cubic kilometres of magma 66 million years ago, which would have had a devastating effect on the global environment. These eruptions probably overlapped the Chicxulub impact and would have accentuated and prolonged its effects on the environment.

While the end of the Cretaceous is marked by this mass extinction, with the loss of the dinosaurs and other important groups such as the ammonites, many other groups of organisms, such as flowering plants and mammals, were relatively unaffected. As with previous extinction events, the aftermath cleared ecological niches to allow new occupants, as life carried on. Among the principal beneficiaries this time were the mammals, including our own ancestors. The Cretaceous is the final period of the Mesozoic and the scene is now set for the third and final era of the stratigraphic column, the Cenozoic, the era of *new life*.

Cenozoic Era

The Cenozoic Era began 65mya and extends to the present. During the course of this era the configuration of continents comes to resemble the pattern we are familiar with today, and the flora and fauna resemble more and more modern forms. The first period of the Cenozoic is the Palaeogene (65–23mya), followed by the Neogene (23–2.6mya). The Palaeogene and Neogene were formerly known as the Tertiary. The high point of global sea level in the Mesozoic was reached in the late Cretaceous, and since then it has steadily declined. Global temperatures in the early Cenozoic were higher than today and began to cool about 50mya, leading to the development of glaciations on the Antarctic continent about 35mya and in the Northern Hemisphere around 3mya in the late Pliocene. Glaciation continued into the Pleistocene, the first epoch of the succeeding Quaternary Period (2.6mya to present). See Figure 5.31.

A feature of the Cenozoic is the formation of a number of the great mountain chains of the modern world. The Alps and Carpathians in southern Europe were formed by the ongoing collision of the African and European plates, while the northward passage of India into southern Asia formed the Himalayas.

Figure 5.30 Cretaceous-Palaeogene boundary at Drumheller, Alberta, Canada.

76

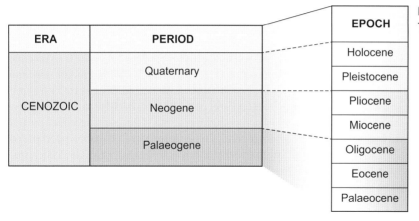

ERA	PERIOD		EPOCH
CENOZOIC	Quaternary		Holocene
			Pleistocene
	Neogene		Pliocene
			Miocene
	Palaeogene		Oligocene
			Eocene
			Palaeocene

Figure 5.31 The sub-division of the Quaternary. The current epoch is the Holocene.

The configuration of the continents by 14mya in the Miocene epoch of the Neogene showed the North and South Atlantic Oceans clearly defined, with India no longer an island but an integral part of Asia, and the Mediterranean Sea still open to the Indian Ocean (see Fig. 5.32).

The Neogene is succeeded by the Quaternary Period, divided into two epochs: the Pleistocene (2.6mya to 11.7kya) and the Holocene (11.7kya to today). The Pleistocene epoch is dominated by a series of glacial and interglacial cycles, principally in the Northern Hemisphere. See Figure 5.8 for the extent

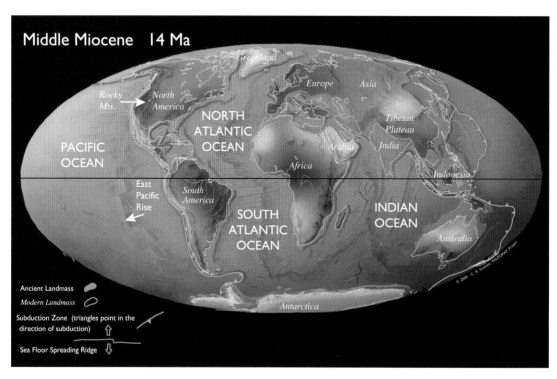

Figure 5.32 Palaeogeography of the Middle Miocene epoch, 14mya.

of glaciations in the northern hemisphere 25 thousand years ago. The effects of glaciations on this scale are widespread and include profound changes in the landscape brought about by erosion and deposition over large areas of the continents affected. Figure 5.33 shows the Aletschgletscher, Switzerland, a typical valley glacier.

In addition, the development of continental ice caps up to 4km thick is the equivalent of a drop in sea level of around 120m. As discussed earlier the weight of ice involved causes isostatic adjustment of the crust where it sags downward and then rebounds when the ice eventually melts. The Pleistocene epoch ended about 12kya, to be followed by the Holocene epoch, which represents the youngest interglacial interval of the Quaternary Period covering the last approximately 12kyr of geological time. It is different from the earlier sub-divisions

of the stratigraphic column because it includes the chronology for *human history* and overlaps with the time frame used by archaeologists. The Holocene therefore covers the evolution and development of *Homo sapiens*. The rise of the human species, with the attendant developments in civilizations and profound changes in life style, such as the Industrial Revolution and a global move to urban living, has had major impacts on the Earth's environment. The accelerated use of fossil fuels since the Industrial Revolution, for example, has impacted the Earth's atmosphere, perhaps with far-reaching consequences. Such has been the influence of the human species on the environment that a new epoch, the Anthropocene, has been proposed to cover the significant impact of *Homo sapiens*. This term is currently under consideration by the International Committee on Stratigraphy for formal adoption as a

77

Figure 5.33 Valley glacier, the Aletschgletscher, Switzerland.
©TravelGretl/Shutterstock

INTERNATIONAL CHRONOSTRATIGRAPHIC CHART

IUGS www.stratigraphy.org   International Commission on Stratigraphy   v 2018/08

Phanerozoic — Cenozoic / Mesozoic

Erathem / Era	System / Period	Series / Epoch	Stage / Age	numerical age (Ma)
Cenozoic	Quaternary	Holocene	Meghalayan (U/L)	present / 0.0042
			Northgrippian (M)	0.0082
			Greenlandian (L)	0.0117
		Pleistocene	Upper	0.126
			Middle	0.781
			Calabrian	1.80
			Gelasian	2.58
	Neogene	Pliocene	Piacenzian	3.600
			Zanclean	5.333
		Miocene	Messinian	7.246
			Tortonian	11.63
			Serravallian	13.82
			Langhian	15.97
			Burdigalian	20.44
			Aquitanian	23.03
	Paleogene	Oligocene	Chattian	27.82
			Rupelian	33.9
		Eocene	Priabonian	37.8
			Bartonian	41.2
			Lutetian	47.8
			Ypresian	56.0
		Paleocene	Thanetian	59.2
			Selandian	61.6
			Danian	66.0
Mesozoic	Cretaceous	Upper	Maastrichtian	72.1 ±0.2
			Campanian	83.6 ±0.2
			Santonian	86.3 ±0.5
			Coniacian	89.8 ±0.3
			Turonian	93.9
			Cenomanian	100.5
		Lower	Albian	~ 113.0
			Aptian	~ 125.0
			Barremian	~ 129.4
			Hauterivian	~ 132.9
			Valanginian	~ 139.8
			Berriasian	~ 145.0

Phanerozoic — Mesozoic / Paleozoic

System / Period	Series / Epoch	Stage / Age	numerical age (Ma)
Jurassic	Upper	Tithonian	~ 145.0
		Kimmeridgian	152.1 ±0.9
		Oxfordian	157.3 ±1.0
	Middle	Callovian	163.5 ±1.0
		Bathonian	166.1 ±1.2
		Bajocian	168.3 ±1.3
		Aalenian	170.3 ±1.4
	Lower	Toarcian	174.1 ±1.0
		Pliensbachian	182.7 ±0.7
		Sinemurian	190.8 ±1.0
		Hettangian	199.3 ±0.3 / 201.3 ±0.2
Triassic	Upper	Rhaetian	~ 208.5
		Norian	~ 227
		Carnian	~ 237
	Middle	Ladinian	~ 242
		Anisian	247.2
	Lower	Olenekian	251.2
		Induan	251.902 ±0.024
Permian	Lopingian	Changhsingian	254.14 ±0.07
		Wuchiapingian	259.1 ±0.5
	Guadalupian	Capitanian	265.1 ±0.4
		Wordian	268.8 ±0.5
		Roadian	272.95 ±0.11
	Cisuralian	Kungurian	283.5 ±0.6
		Artinskian	290.1 ±0.26
		Sakmarian	293.52 ±0.17
		Asselian	298.9 ±0.15
Carboniferous — Pennsylvanian	Upper	Gzhelian	303.7 ±0.1
		Kasimovian	307.0 ±0.1
	Middle	Moscovian	315.2 ±0.2
	Lower	Bashkirian	323.2 ±0.4
Carboniferous — Mississippian	Upper	Serpukhovian	330.9 ±0.2
	Middle	Visean	346.7 ±0.4
	Lower	Tournaisian	358.9 ±0.4

Phanerozoic — Paleozoic

System / Period	Series / Epoch	Stage / Age	numerical age (Ma)
Devonian	Upper	Famennian	358.9 ±0.4
		Frasnian	372.2 ±1.6
	Middle	Givetian	382.7 ±1.6
		Eifelian	387.7 ±0.8
	Lower	Emsian	393.3 ±1.2
		Pragian	407.6 ±2.6
		Lochkovian	410.8 ±2.8
Silurian	Pridoli		419.2 ±3.2
	Ludlow	Ludfordian	423.0 ±2.3
		Gorstian	425.6 ±0.9
	Wenlock	Homerian	427.4 ±0.5
		Sheinwoodian	430.5 ±0.7
	Llandovery	Telychian	433.4 ±0.8
		Aeronian	438.5 ±1.1
		Rhuddanian	440.8 ±1.2
Ordovician	Upper	Hirnantian	443.8 ±1.5 / 445.2 ±1.4
		Katian	453.0 ±0.7
		Sandbian	458.4 ±0.9
	Middle	Darriwilian	467.3 ±1.1
		Dapingian	470.0 ±1.4
	Lower	Floian	477.7 ±1.4
		Tremadocian	485.4 ±1.9
Cambrian	Furongian	Stage 10	~ 489.5
		Jiangshanian	~ 494
		Paibian	~ 497
	Miaolingian	Guzhangian	~ 500.5
		Drumian	~ 504.5
		Wuliuan	~ 509
	Series 2	Stage 4	~ 514
		Stage 3	~ 521
	Terreneuvian	Stage 2	~ 529
		Fortunian	541.0 ±1.0

Precambrian

Eonothem / Eon	Erathem / Era	System / Period	numerical age (Ma)
Proterozoic	Neoproterozoic	Ediacaran	541.0 ±1.0 / ~ 635
		Cryogenian	~ 720
		Tonian	1000
	Mesoproterozoic	Stenian	1200
		Ectasian	1400
		Calymmian	1600
	Paleoproterozoic	Statherian	1800
		Orosirian	2050
		Rhyacian	2300
		Siderian	2500
Archean	Neoarchean		2800
	Mesoarchean		3200
	Paleoarchean		3600
	Eoarchean		4000
Hadean			~ 4600

Units of all ranks are in the process of being defined by Global Boundary Stratotype Section and Points (GSSP) for their lower boundaries, including those of the Archean and Proterozoic, long defined by Global Standard Stratigraphic Ages (GSSA). Charts and detailed information on ratified GSSPs are available at the website http://www.stratigraphy.org. The URL to this chart is found below.

Numerical ages are subject to revision and do not define units in the Phanerozoic and the Ediacaran; only GSSPs do. For boundaries in the Phanerozoic without ratified GSSPs or without constrained numerical ages, an approximate numerical age (~) is provided.

Ratified Subseries/Subepochs are abbreviated as U/L (Upper/Late), M (Middle) and L/E (Lower/Early). Numerical ages for all systems except Quaternary, upper Paleogene, Cretaceous, Triassic, Permian and Precambrian are taken from 'A Geologic Time Scale 2012' by Gradstein et al. (2012); those for the Quaternary, upper Paleogene, Cretaceous, Triassic, Permian and Precambrian are provided by the relevant ICS subcommissions.

Colouring follows the Commission for the Geological Map of the World (http://www.ccgm.org)

Chart drafted by K.M. Cohen, D.A.T. Harper, P.L. Gibbard, J.-X. Fan (c) International Commission on Stratigraphy, August 2018

CCGM / CGMW

To cite: Cohen, K.M., Finney, S.C., Gibbard, P.L. & Fan, J.-X. (2013; updated) The ICS International Chronostratigraphic Chart. Episodes 36: 199-204

URL: http://www.stratigraphy.org/ICSchart/ChronostratChart2018-08.pdf

Figure 6.1 International Chronostratigraphic Chart. *Reproduced by permission ©ICS International Commission on Stratigraphy [2018].*

The International Chronostratigraphic Chart (The ICS Chart) is a hierarchy of chronostratigraphic units (Eonothems, Erathems, Systems, Series and Stages) on which geochronological units (Eons, Eras, Periods, Epochs and Ages) are based.

In stratigraphy an eonothem is the totality of rock strata laid down in the stratigraphic record deposited during a certain eon of the continuous geologic timescale. An erathem is the total stratigraphic unit deposited during a certain corresponding span of time during an era in the geologic timescale.

A system consists of all the rocks formed during a period, and systems and periods have the same names, so the rocks of the Cambrian *System* were deposited during the 54myr of the Cambrian *Period*. Systems can be divided into shorter chronostratigraphic units – series, stages and chronozones (see table 6.1). A **series** consists of the rocks formed during

Table 6.1 Units in geochronology and chronostratigraphy (after Cohen *et al.*, 2015).

Segments of rock (strata) in chronostratigraphy	Timespans in geochronology	Notes to geochronological units
Eonothem	Eon	4 total, half a billion years or more
Erathem	Era	10 defined, several hundred million years
System	Period	22 defined, tens to ~one hundred million years
Series	Epoch	34 defined, tens of millions of years
Stage	Age	99 defined, millions of years
Chronozone	Chron	Subdivision of an age, not used by the ICS timescale

an **epoch**, a **stage** is formed during an age, and the shortest time unit, the chron, accounts for the formation of a chronozone.

The hierarchy of terms used for lithostratigraphic units from largest to smallest is: Supergroup, Group, Formation, Member and Bed. For the purposes of geological mapping, which after all is the fundamental process that underpins the whole of stratigraphy, it is possible to recognize **groups**, **formations** and **members**. Formations are the primary unit in lithostratigraphy, and as well as stratigraphic position they share some distinctive lithological features such as mineral composition, texture, sedimentary structures and fossil content. They are units large enough to be mapped. The age of a formation is not necessarily the same wherever it is identified. A number of formations may be combined into a group, or subdivided into members. Members may be subdivided into **beds** and groups may be combined to form Supergroups.

The primary objective of the ICS is the precise definition of a global standard set of time-correlative units – Systems, Series and Stages – for stratigraphic successions across the globe. These time-based units in turn are the basis for the Periods, Epochs and Ages of the Geological Timescale when calibrated numerical ages for the unit boundaries are added. The setting of an internationally recognized global standard is essential for the communication of geological knowledge, not only to professionals in the Earth Science community but also for decision-makers in government and industry and those members of the general public who may be interested in Earth history. One example of the importance of stratigraphy is the current concern about climate change and the degree to which it is due to human activity. Such changes can only be fully understood by a study of similar climate variations in the geological past.

The 'Golden Spike' in stratigraphy

In 1869 at Promontory Summit in Utah Territory, Leland Stanford, chairman of the Central Pacific Railway, drove in a golden spike to join the rails to mark the completion of the First Transcontinental Railroad across the United States as the Central Pacific and Union Pacific railroads met. Stanford went on to found Stanford University in California, the present location of the Golden Spike.

Since then the concept of a Golden Spike as a marker has been used in a variety of contexts. For many years it has been put forward as a way of solving problems in stratigraphy involving disputed boundaries between subdivisions of the stratigraphic column.

The ICS use **Global Boundary Stratotype Section and Points (GSSP)** as internationally agreed reference points on a stratigraphic section to define the lower boundary of a *stage* on the geologic timescale. Most but not all GSSPs are based on palaeontological changes and are usually described in the context of transitions between faunal stages. Since 1977, when the procedure was initiated, 67 of the 100 stages in the Phanerozoic that need a GSSP have been formally defined. The International Chronostratigraphic Chart (Fig. 6.1) is organized into four main columns, three of which represent the Phanerozoic Eon. Each main column further subdivided to show the relevant era, system, series and stage. So, for example, the left-hand column shows the upper part

86

of the Phanerozoic Eon with the Cretaceous, Palaeogene, Neogene and Quaternary systems. These systems are divided into series and stages. At their lowest subdivision level the three main columns comprise 100 stages. The fourth main column on the right of the diagram shows the Precambrian, subdivided to system level. Longer stages have thicker intervals in the columns. Units of all ranks are in the process of being defined by GSSPs for their lower boundaries, including those of the Archaean and Proterozoic. The latter has long been defined by **Global Standard Stratigraphic Ages (GSSA)**. On the chart the status of each GSSP is marked by a small golden-spike icon at the base of the relevant division they define. For each GSSA in the Proterozoic and Archaean there is a small clock icon. The Ediacaran in the Neo-Proterozoic is defined by a golden spike.

Criteria for adoption as a GSSP by the ICS

For any geological section to be considered as a potential GSSP it must fulfil a set of criteria. These can be defined as:

• A GSSP must define the lower boundary of a geological stage;
• The lower boundary must be defined by a primary marker, usually the first appearance of a fossil species;
• There should also be secondary markers such as other fossils, chemical evidence or geomagnetic evidence;
• The horizon in which the marker appears should have minerals capable of being radiometrically dated;
• The marker should have regional and global correlation in outcrops of the same age;
• The marker should be independent of facies;
• The marker should have adequate thickness;
• The outcrop should be unaffected by tectonic or sedimentary movements, or by metamorphism;
• The outcrop must be in accessible terrain and located where it can be visited quickly;
• The outcrop must be extensive enough to allow repeated sampling and be open to researchers of all nationalities.

Reference: Pirgen Remane, Michael G. Bassett, John W. Cowie, Klaus H. Gohrbandt, H. Richard Lane, Olaf Michelsen and Wang Noiwen, with the cooperation of members of ICS. Revised guidelines for the establishment of global chronostratigraphic standards by the International Commission on Stratigraphy (ICS).

Procedure for the submission of a GSSP

A submission to ICS for a named boundary must give information indicating the exact location of the section on a detailed topographic and geological map. Details should be given of the section, including the bed in which the boundary point is defined, and the key-levels for all physical and biostratigraphic markers. Further information should indicate elements such as motivation for the choice of boundary level and a discussion of all markers used in the determination of the boundary level. For full details of submission procedures see relevant references and website.

Following acceptance of the submission within the guidelines, the Chairperson of ICS will arrange a vote by the full Commission within a period of not more than 60 days.

Case Study 1: the Ediacaran Period – a new addition to the geological timescale

Palaeontologists in the nineteenth century, when many of the periods of the geological timescale were being designated, believed that the trilobite and brachiopod fauna of the early Cambrian period represented the earliest known life forms. Since then the fossil record of life on Earth has been extended back to 3.5byr. Many of these fossils are microscopic forms; however, in what is now recognized as the Ediacaran, macroscopic fossils of soft-bodied animals have been found, such as *Cyclomedusa plana* (see Fig. 4.8), thought to be similar to modern-day jellyfish. Fossils of the Ediacaran Period, duration 635–541bya, give us a strong indication of life on Earth at the end of the Proterozoic Era.

In 2004, in accordance with the protocols outlined above and following the procedures of the International Commission on Stratigraphy, the International Union of Geological Sciences has approved a new addition to the geological timescale, the Ediacaran Period. This represents the first Proterozoic period to be recognized on the basis of chronostratigraphic criteria and the first internationally ratified, chronostratigraphically defined period of any age to be introduced in 120 years. The significance of the Ediacaran period is its contained fossil fauna. The base of the Ediacaran Period is defined by a Global Stratotype Section and Point (GSSP) at the base of the Nuccaleena Formation at Enorma Creek in the Flinders ranges of South Australia (see (see figures 6.2 and 6.3).

Figure 6.2 GSSP for the base of the Ediacaran Period, Australia.

Figure 6.3 Close up of the 'golden spike'.

The rock is a **cap carbonate**, lying directly above glacial diamictites or boulder clays of the Marinoan glaciation, a period of worldwide glaciations during the Cryogenian period of the Neoproterozoic. Cap carbonates are continuous layers of limestone that sharply overlie Neoproterozoic glacial

deposits. Their formation is likely to be associated with the rapid decay of the Marinoan ice sheets and changes in ocean chemistry due to CO_2 build-up. The significance of cap carbonates is that they form distinctive marker horizons on a global scale.

The top of the Ediacaran is defined by the initial GSSP of the Cambrian Period. Thus the new Ediacaran Period consists of a distinctive unit of geological time bounded above and below by similarly distinctive episodes.

Case Study 2: the Ludlow Bone Bed and the Silurian–Devonian junction

In chapter 2 dealing with the historical development of stratigraphy, the important role played by stratigraphers based in Britain in the recognition and designation of the periods and systems of the stratigraphic column was discussed. All six periods of the Palaeozoic Era, from Cambrian to Permian, were originally proposed by British geologists; the term Cambrian refers to the Latin name for Wales, and the Ordovician and Silurian were named after ancient Welsh tribes. As detailed in chapter 2, Sir Roderick Murchison established the Silurian Period, and with Adam Sedgwick in 1839 designated the succeeding Devonian Period, named after the English county.

Some years earlier in 1832, near Ludlow in the Welsh Borders, Murchison had been shown a thin layer of dark sand containing remains of fish scales and plant debris, and which was therefore a terrestrial deposit. Lying immediately below this bed was the Ludlow Series, 350m of siltstones and limestones, which had been deposited in a warm, shallow sea. The so-called Ludlow Bone Bed therefore represents a significant change in geological environment (see Fig. 6.4).

At the time, this bed was believed to be the earliest occurrence of life on land, so Murchison took it as the boundary between the Silurian and Devonian systems. However, scientific knowledge is constantly changing, and as more became known about the rocks on the Silurian–Devonian boundary across the world, then it became clear the Ludlow Bone Bed was not the best marker for this junction. In the next century the top of the Silurian System was moved up from the Bone Bed, with the overlying rocks being designated part of the *Pridoli* Series, marked by the first occurrence of a graptolite species *Monograptus parultimus*. See Figure 6.5 from the relevant part of the International Chronostratigraphic Chart.

7 The application of stratigraphy

Figure 7.1, an image of a drilling rig in the oil and gas fields of the North Sea is a familiar one now to most people. Platforms such as these play an essential role in the exploration process for oil and gas reserves. Exploratory holes are drilled to assess the potential existence of a hydrocarbon reserve and then changed to a production role if the exploration proves successful. **Borehole logging**, or well logging, is the practice of making a well log, a detailed record, of the sequence of rocks encountered as the borehole penetrates the geological succession. This log can be based on a number of parameters, depending on the reasons for drilling the borehole. It may be based on the visual inspection of the samples brought to the surface or on measurements made 'downhole' by various geophysical instruments lowered after the hole has been drilled. Coring is the technique whereby an actual sample is removed using a specialized drill-bit.

Mud logging is the procedure where a mud log is made by examining rock cuttings brought to the surface by the drilling mud that circulates in the drill hole to cool and lubricate the drilling. The mud log will include parameters such as the rate of penetration, the formation gas, the lithology and the temperature of the drilling fluid.

As well as drilling for hydrocarbons, boreholes are used for other purposes such as groundwater assessment, geotechnical investigations and geothermal and mineral exploration.

Wireline logging

The term 'wireline logging' refers to the technique, widely used in the oil and gas industry, whereby one or more instruments are lowered on the end of a wireline into a borehole. A variety of sensors can then be used to record the properties of

Figure 7.1 Deepsea Delta semi-submersible drilling rig in the North Sea.
©Dabarti CGI/Shutterstock

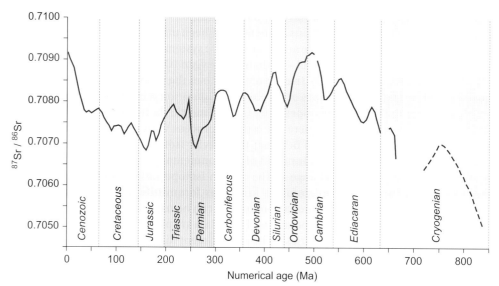

Figure 7.8 Marine sediment Sr-isotope curve for the Phanerozoic. After McArthur *et al*, 2001.

distribution and stratigraphy of subsurface rock successions is exploration seismology. The technique can be carried out remotely, that is, without having to drill a borehole. Such techniques are used to erect a seismic stratigraphy using explosive or vibrational wave sources and an array of detectors (geophones) to examine the time–distance–travel relationships through the rock succession. While developed and used extensively in the petroleum industry, shallow reflection seismic techniques are now also used to investigate environmental problems involving, for example, water table conditions and to identify and monitor zones of subsurface contamination.

The illustration in Figure 7.9 shows a commonly used technique developed to explore the stratigraphy below the sea bed in an area being investigated for potential oil or gas deposits offshore. A source of seismic waves or vibrations, powered by compressed air, creates sound waves. These waves propagate outwards from the ship and reflect off the undersea rock layers. The reflected waves are recorded by sensors known as hydrophones, towed behind the ship. The differences in the ways various rock types transmit the seismic waves allows the recognition of the sedimentary rock layers and also the existence of any *reservoir* rocks containing water, oil or gas.

If the exploration site is on land, a variation on the marine method is to use a vibration source, usually based on a large truck, to generate seismic waves by shaking the ground. The returning waves are recorded by an array of geophones, just as in the marine set-up.

In the hydrocarbons and minerals industries seismic exploration is used in the search for commercially viable subsurface deposits of oil and gas and other minerals using techniques such as those outlined above.

The seismic waves generated artificially can be analysed in different ways. When they cross a geological interface, generally marked by materials of different density, part of the wave energy is refracted, or bent, into the underlying layer, and the remainder is reflected back at the same angle as it arrived. The reflection and refraction of seismic waves at density changes are covered by exactly the same laws as those that govern the passage of light rays. When a seismic wave passes from one medium to another there is a change in the velocity of the wave. The change in physical properties will cause the wave to bend or refract, while a portion of the wave energy is bounced back or reflected. The behaviour of the refracted wave is governed by Snell's Law:

$$\text{Snell's Law: } \sin i / \sin r = V_1/V_2$$

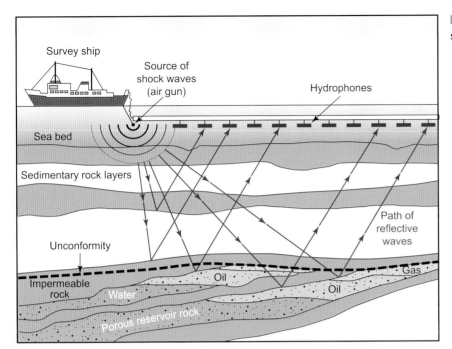

Figure 7.9 Illustrates the basic principles of seismic data acquisition in a marine setting.

where angle i is the angle of incidence of the ray and angle r is the angle of refraction and V_1 and V_2 are the velocities of wave propagation in the two layers.

Since density affects the velocity with which the wave passes through the rock, density changes mean that arrival times of the various types of wave differ. The waves slow down entering a denser substance and speed up going into a less dense medium. Analysis of these differences in arrival times allows the construction of tomographic images of the detected subsurface structures such as structural traps and potential reservoir rocks. Seismic tomography is a technique for imaging the subsurface of the Earth using seismic waves (see Figs. 7.10 and 7.11).

The depth of penetration of the seismic waves depends on the scale of the energy source and the number of geophones in the array, but can vary from a few metres to several kilometres. Seismic reflection profiling is the principal method used in the hydrocarbon industry to explore for oil and gas, and provides the basis for modern seismic stratigraphy. The main limitations of the use of reflection techniques are its

restriction to depths greater than 15m and its higher cost when compared to refraction methods. The use of seismic reflection is particularly suited to marine conditions or other environments such as rivers and lakes. Figure 7.12 shows typical reflector data collected in the Gulf of Mexico by the US Geological Survey in a study investigating the occurrence of gas hydrate, a crystalline form of water and methane, in the sea-bed sediments.

Seismic refraction is generally applicable only in environments where the seismic velocities of the layers increase with depth, otherwise incorrect results may be obtained. In addition, seismic refraction requires geophone arrays with lengths four to five times the distance to the interface of interest, meaning that the technique is normally limited to depths less than 30m. Greater depths can only be achieved by larger explosive sources and more extensive geophone arrays. In situations where both techniques are applicable, seismic reflection will normally have better resolution, but since it is considerably more expensive than using refraction techniques (three to five times more) it may come down to a budget decision.

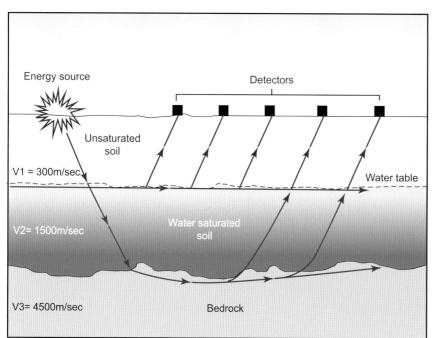

Figure 7.10 Collection of seismic refraction data.

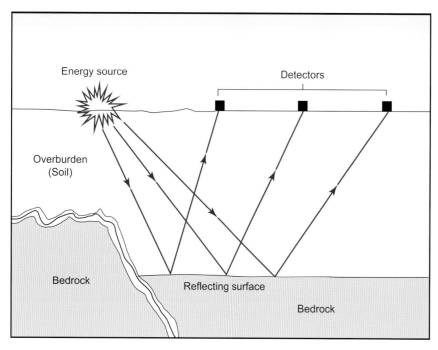

Figure 7.11 Collection of seismic reflection data.

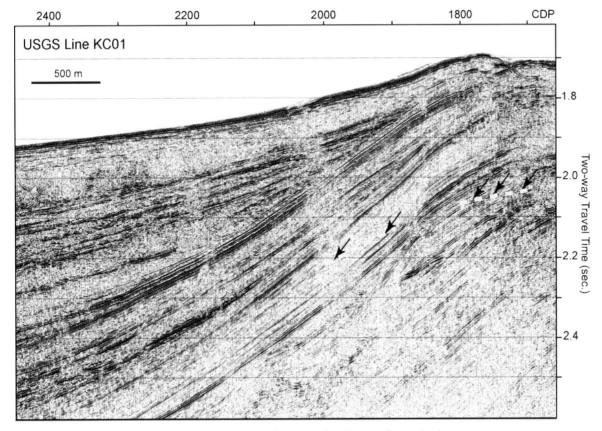

Figure 7.12 Seismic reflection data from the Gulf of Mexico showing a reflector horizon.

Sequence stratigraphy

The most radical change in thinking about sedimentary rocks in the last few decades, **sequence stratigraphy** is that branch of stratigraphy that attempts to subdivide sedimentary deposits into units bounded by unconformities. It was developed in the mid-1970s by geologists working in the oil industry as an aid for accurate correlation and integration between measured sections in outcrop, wells and seismic data. Sequence stratigraphy explains the accumulation of sediments in response to relative rates of change in sea level. It is a *chronostratigraphic* methodology providing a practical means of subdividing the stratigraphic record. Using the technique allows the identification and *prediction* of facies and records the time-transgressive nature of *lithostratigraphic* units.

As explained in chapter 5, eustatic (or global or absolute) sea level is measured relative to a fixed point, the centre of the Earth. Relative sea level is measured against the level of the continental crust, so variation in sea level can therefore occur by local vertical land movements and/or eustatic/absolute sea level changes. The contributions by Nicolas Steno and Johannnes Walther to sedimentary interpretation, and therefore to sequence stratigraphy, were hugely influential. Steno's principle of original horizontality showed that not only are the oldest sediments at the bottom of a succession, but deposition will fill a basal irregular surface enclosed above by a smooth surface (see Fig. 7.13). His principal of lateral continuity stated that contemporaneous layers of sediment are continuous until they pinch out, or a barrier prevents their further spread during

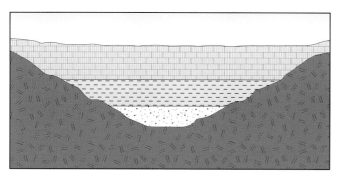

Figure 7.13 Basal irregular surface filled by sediments showing original horizontality.

deposition. Following on from this premise, Walther recognized that as depositional settings change their lateral position and build successions, so the sedimentary facies of depositional settings succeed one another as a vertical sequence. Figure 3.16 shows the change in sedimentary facies in the case of marine transgression and regression, and Walther's Law deals with the relationship between vertical successions and lateral sequences (see Fig. 3.18). A sequence is defined as a relatively conformable succession of genetically related transgressive-regressive strata bounded at the upper surfaces and base by unconformities and their correlative conformities. Sequence stratigraphy builds on these fundamental concepts of stratigraphy to divide a sequence of sediments into time-related units that can then be used for correlation and facies prediction. Sequence stratigraphy deals with the order, or sequence, in which depositionally related successions of strata were laid down in the available space or accommodation.

Accommodation is defined as 'the space available for potential sediment accumulation', and there are a number of controlling factors. This space is the combined product of movement of:
1 The sea surface: global sea level measured from a datum such as the centre of the Earth;
2 The sea floor (tectonics);
3 Changes in rates of sediment accumulation. This sediment supply will be siliciclastic sediment from land and/or carbonate sediment in the depositional basin (see Fig. 7.14).

Siliciclastic rocks are **clastic** or fragmentary sedimentary rocks composed almost entirely of quartz or other silicate minerals. Carbonate rocks are mostly limestones, composed of the mineral calcite, $CaCO_3$.

Fundamentally there is a balance between the relative rate of increase of accommodation making space for more sedimentation *versus* the rate of supply of sediment to fill that accommodation.

The layering of sedimentary rocks is expressed as sets of simple to complex arrangements of beds, and a wide variety of different sedimentary facies. Each of the factors controlling the precise nature of sequences is capable of a wide range of values. This can lead to wide variety in the resulting deposits. Many geologists using stratigraphy use subsurface data rather than outcrop data, particularly in the oil industry. The principal source of data is from exploration seismology, where the identification of different reflecting layers is used to interpret the stratigraphic order.

The chronostratigraphy of sedimentary rocks records changes in their character through geological time. These changes may be shown in graphical form as geological cross-sections, as chronostratigraphic correlation charts,

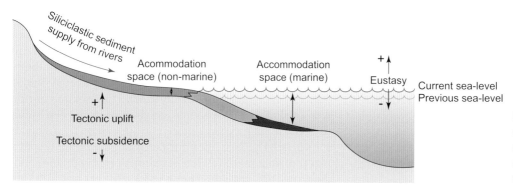

Figure 7.14 Accommodation space available for sediment accumulation based on changes in eustatic sea level, tectonics and sediment input.

or Wheeler diagrams. An example of a Wheeler diagram is shown below (Fig. 7.15A). It is a stratigraphic summary chart on which geological time is plotted as the vertical scale, and distance across the area of interest as the horizontal scale, marked by the geographic position of shot points (SP). A reflector is any boundary between two layers with different densities and seismic wave speed. It is possible to include a range of stratigraphic information including the horizontal distribution of the contemporaneous sedimentary layers of a sequence, as well as any significant breaks in sedimentation. Details of interpretation of these diagrams will be discussed in later sections.

- Onlap: by which packages of strata with relatively gentle dip are banked up against an older package of strata with greater dip;
- Downlap: by which a package of inclined strata terminates downward against a less steeply dipping package of strata;
- Toplap: by which inclined strata terminate upwards against a less steeply dipping surface.

Truncations marked here usually involve erosion and are unconformities (see Fig. 7.15B).

This chronostratigraphic correlation is distinct from the geochronology or geological age of strata. The discipline of sequence stratigraphy provides a tool used to interpret the

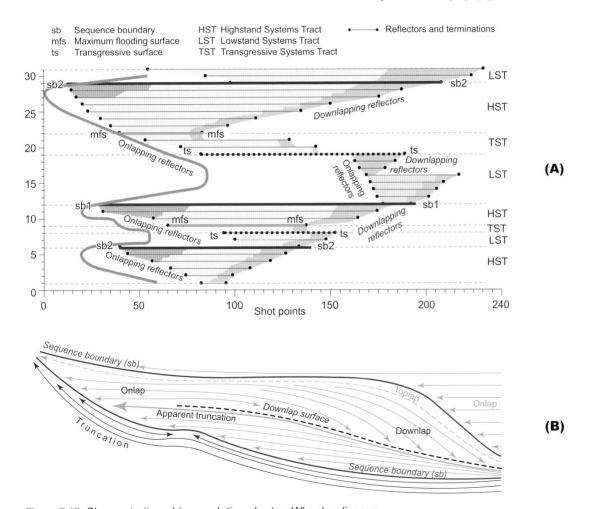

Figure 7.15 Chronostratigraphic correlation chart or Wheeler diagram.

depositional origin and predict the heterogeneity, extent and character of the lithofacies. The term 'lithofacies' refers to a facies characterized by particular lithological features; it is a subdivision of a designated stratigraphic unit that is capable of being mapped. Sequence stratigraphy combines the framework of major depositional and erosional surfaces bounding these successions of strata and the geometry that successive contemporaneous strata have, following their accumulation within that framework.

Sea-level cycles

Sequence stratigraphy is so-named because it refers to sedimentary deposits that are laid down in cycles. Each depositional sequence records one cycle of relative sea-level rise and fall. A sequence begins with a slow rise in sea level following a fall and continues until the next fall in sea level. A sequence is bounded by chronostratigraphic surfaces that include

unconformities, formed when the sea level was falling, and flooding surfaces formed during relative sea-level rise. The unconformities include features such as incised river valleys, soil processes and vegetation cover. These surfaces, and the geometries of the sediments they enclose, are combined to interpret the depositional setting of clastic and carbonate sediments. These settings include continental, marginal marine, basin margins and downslope settings of basins.

Seismic evidence is used to find evidence of unconformities where transgressing seas have resulted in the laying down of younger beds over older beds as seas deepen, allowing the drawing of sea-level curves showing the variation of sea level with time (see Fig. 7.16). Pioneered by Peter Vail of the Exxon oil company in the 1970s, these curves are used in the evaluation of potential hydrocarbon reservoirs.

When looked at over the duration of geological time, the changes in sea level show cycles at a number of orders. A

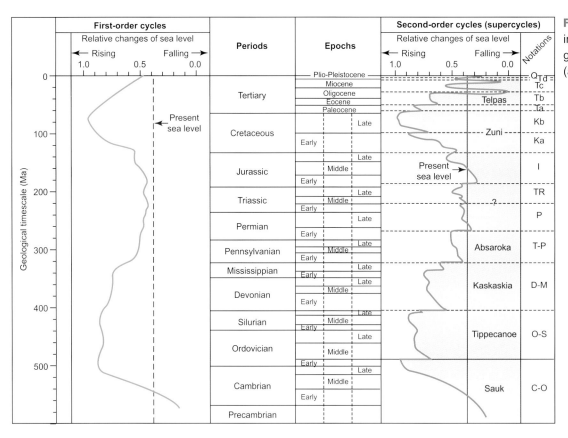

Figure 7.16 Changes in sea level through geological time (after Vail et al 1977).

broad cycle lasting hundreds of millions of years represents a *first-order cycle* linked to major plate-tectonic events such as continental breakup, while a *second-order cycle* of frequent transgressive and regressive events occurs with a periodicity of tens of millions of years and is linked to astronomical cycles. The *third-order cycle* has a periodicity of around 100kyr and is linked to the oscillation of the Earth's orbit from elliptical to circular. The *fourth-order cycle* of about 40kyr is explained by the rate at which the Earth's inclination to the Sun varies, and the *fifth-order cycle*, the smallest scale of these cycles, is approximately 20kyr and corresponds to the rate of precession of the Earth's rotational axis. Precession is the slow movement of a spinning body around another axis under the influence of another force such as gravity. It is best demonstrated by the circle slowly traced out by a spinning gyroscope. The third-, fourth- and fifth-order cycles can be explained by the Milankovitch cycle. This is a cyclical movement related to the Earth's orbit around the Sun and involves the eccentricity of the Earth's orbit, the tilt of its axis, and precession as described above. Table 7.1 lists five orders of sea-level cycles and defines them by duration. Since there are several orders of sea-level cyclicity ranging from thousands of years to tens or hundreds of millions of years, sedimentary successions may be deposited during more than one order of sea-level cycle.

The hierarchy of surfaces used for subdivision in sequence stratigraphy is:

• Sequences
• Systems tract
• Parasequences and/or cycles
• Bedsets
• Beds

The genetic relationship between the sequences, systems, bedsets and beds is a product of changes in sediment accommodation. Beds are layers of sedimentary rocks bounded above and below by bedding surfaces. Bedding surfaces are produced during periods of non-deposition, or abrupt changes in the conditions of deposition, including erosion. Since bedding surfaces are synchronous when traced laterally, they are therefore time-stratigraphic units.

A bedset consists of two or more superimposed beds having the same composition, texture and sedimentary structures. A bedset is a record of deposition in an environment, formed by certain depositional processes, all of which make up the sedimentary facies. Bedset boundaries form over a longer time period than beds, and usually have a greater lateral extent than beds.

A parasequence is a relatively conformable, genetically related succession of beds and bedsets bounded by **marine flooding surfaces** and their correlative surfaces. A marine flooding surface separates younger from older strata, across which there is evidence of an abrupt increase in water depth. This may be accompanied by minor submarine erosion or non-deposition, with a minor hiatus indicated.

Sequence boundaries

Sequence boundaries are the most significant surfaces in sequence stratigraphy. Sequences are generated by cycles of change in sediment accommodation or sediment supply that form similar sequence stratigraphic surfaces. Consider the events during a complete sea-level cycle in an environment consisting of a broad shelf area bounded by land in one direction and dropping off into a marine basin on the other. As the sea level falls (a regression) the siliciclastic sediment from the land, or the carbonate sediment derived *in situ*, is deposited in progressively shallow water, producing a *shallowing-upward* (becoming coarser upwards) succession. As sea level continues to fall, the depositional area may become

Table 7.1 Sea-level cycle order.

Cycle order	Nomenclature	Thickness range (m)	Aerial extent (km²)	Duration (Ma)	
				Range	Mode
1st	Megasequence	300m +	Global	50–100+	80
2nd	Supersequence	150–1500+	Regional	5–50	10
3rd	Sequence	150–450	1300–130,000	0.5–5	1
4th	Parasequence Set	5–250	50–5000	0.1–0.5	0.45
5th	Parasequence	5–60	50–5000	0.01–0.1	0.04

emergent and an unconformity is formed. By convention this surface marks the **sequence boundary**. As sea level rises again (a transgression) the depositional area is progressively flooded, and sediment deposited during this phase forms a *deepening-upward* (becoming finer upwards) succession. The transgression culminates in the **maximum flooding surface**. As maximum flooding approaches, rates of sedimentation on the shelf increase and the cycle repeats itself.

System tracts are distinct packages of sediment that are deposited during specific phases of the sea-level cycle. They are suites of co-existing depositional systems such as coastal plains, continental shelves and submarine fans within a sequence. These units are represented in the rock record as three-dimensional facies assemblages. It is possible to recognize four systems tracts within the overall depositional systems that link sea-level changes with sediment supply and the rate of subsidence of the sedimentary basin.

As shown in table 7.2, a complete sequence begins at the Lowstand Systems Tract boundary and ends at Falling-stage Systems Tract boundary. This complete sequence consists of four tracts:

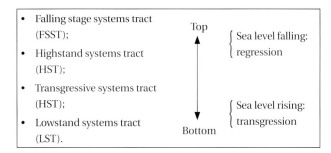

Figure 7.17 shows the sequence of events in the Four Systems Tract Model (after Willis and Fitris, 2012).

A *lowstand systems tract* (Fig. 7.17A) forms during the early stage of the sea-level curve when the rate of sedimentation

Table 7.2 The Four Systems Tract Model.

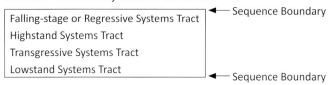

outpaces the rate of sea-level rise. It is bounded by a sequence boundary comprising a subaerial unconformity at the base and the transgressive surface of the tract above.

A *transgressive systems tract* (Fig. 7.17B) forms when the rate of sedimentation is outpaced by the rate of sea-level rise in the sea-level curves. It is bounded by the transgressive surface at its base and the maximum flooding surface at the top.

A *highstand systems tract* (Fig. 7.17C) forms when the rate of sea-level rise drops below the sedimentation rate. It is bounded by the maximum flooding surface at the base and by an erosion surface at the top created as relative sea level begins to fall.

A *falling-stage systems tract* (Fig. 7.17D) forms when the relative sea level is falling and eventually exceeds the rate of sedimentation. It is bounded by an erosion surface at its base and a subaerial unconformity at the top.

The lowstand tract and the transgressive tract are created during transgression of the sea while the highstand to falling-stage tracts occur during regression of the sea. The 4 systems tract model is cyclical, based on the change from a transgressive or deepening sea to a regressive or shallowing sea.

Figure 7.17A shows the maximum state of subaerial exposure and erosion; diagram 7.17C shows the maximum state of flooding. Stages (B) and (D) are transitional between these two extremes.

The concept of systems tracts resulting from transgression and regression events can be related to sea level around Britain and Ireland during the last glacial period. It is estimated that during the Last Glacial Maximum around 20kya, sea level was as much as 125m lower than present levels. This can be regarded as a lowstand, with various land bridges existing between Britain and Ireland and between Britain and Europe. There has been a gradual rise in eustatic sea level since that period due to the progressive melting of the large continental ice sheets. Isostatic rebound has added to this rise in proportion to the degree of crustal depression resulting from ice overburden. At present, sea level is relatively high compared with its position during most of the Quaternary, so modern sea level can be regarded as a highstand tract.

Parasequence stacking
In order to establish the depositional setting of the sedimentary section, sequence stratigraphy uses the geometric arrangement of the sediment fill, particularly the vertical succession

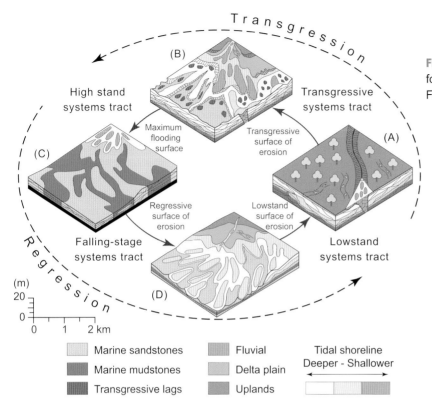

Figure 7.17 Sequence of events in the four systems tract model (after Willis and Fitris, 2012).

of sedimentary facies geometries and their enveloping surfaces, which form the **parasequence stacking pattern**.

The term 'parasequence' is used for the shortest sea-level cycle and produces thicknesses of a few metres. In most cases there will be a series of parasequences, and these sets of successive parasequences are stacked in characteristic ways depending on their position within the sea-level cycle. During the Highstand Systems Tract (HST) phase of the cycle, relative sea level is slowly rising and sediment supply is greater than the rise and drives a seaward building of the coast. This process is called progradational parasequence stacking (see Fig. 7.18). Progradation is the lateral outbuilding of strata in a seaward

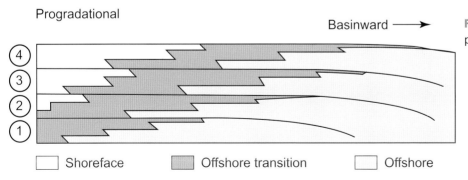

Figure 7.18 Progradational stacking of parasequences.

108

direction. It occurs as the result of a sea-level rise accompanied by a high sediment flux, causing a regression. This usually happens during the late stages of the development of a highstand systems tract and/or during a falling stage systems tract.

In outcrop the sequence can be recognized by the progressive appearance of shallow-water facies upward and is bounded by a flooding surface, across which water depth abruptly increases.

The highstand systems tract (HST) is bounded by the maximum flooding surface at the base (MFS) and an unconformable boundary above. The maximum flooding surface is a surface of deposition at the time when the shoreline is at its maximum landward position (i.e. the time of maximum transgression).

A falling stage systems tract (FSST), sometimes referred to as a regressive systems tract, forms when the relative sea level is falling and eventually exceeds the rate of sedimentation. This creates a condition of *forced regression* in which the coast is forced to build seawards. The fall in sea level leads to erosion, and the resulting erosion surface is the basal surface of forced regression and the beginning of the FSST. During this stage, a river may cut incised valleys on what was previously a marine shelf. The FSST is bounded by a subaerial unconformity at its top.

A lowstand systems tract (LST) forms when the rate of sedimentation is greater than the rate of sea-level rise during the early stage of the sea-level curve. It is bounded by an aerial unconformity at the base, which is the system boundary, and represents the maximum extent of subaerial exposure and erosion. During the LST the incised valleys at sea level begin to flood and become sites of estuaries, which advance landwards as sea level continues to rise, forming the transgressive surface that marks the first of the flooding surfaces that separate the LST from the next phase, the transgressive systems tract (TST).

Retrogradational parasequence stacking

The transgressive system tract forms when the supply of sediment is outpaced by the rate of sea-level rise, causing retrogradational parasequence stacking. Retrogradation is the movement of the coastline landward in response to a transgression. This can occur during a sea-level rise with low sediment flux. Retrogradational stacking patterns refer to patterns of parasequences in which each parasequence is shifted landwards relative to the preceding parasequence (see Fig. 7.19).

A retrogradational parasequence set is one in which successively younger parasequences are deposited farther landward in a backstepping pattern. Overall, the rate of deposition is less that the rate of accommodation. Such retrogradational stacking is characterized by well-developed flooding surfaces, and is typical of the TST. Estuaries are commonly developed during the TST as the valleys cut during the preceding FSST and LST are flooded. In outcrop a retrogradational parasequence set can be recognized by the progressive appearance upwards of deeper water facies within the parasequence set.

Eventually the rate of eustatic rise will slow and be outpaced by the rate of sedimentation, and progradational stacking in a renewed HST will occur. The change from retrogradational stacking in the TST to progradational stacking in the HST corresponds to the deepest water depths in a sequence, the maximum flooding surface (MFS).

The end of the depositional sequence is marked by the return of a fall in sea level and the formation of the FSST.

Aggradational parasequence stacking

If the long-term rate of accommodation closely matches the long-term rate of sedimentation, then aggradational parasequence stacking occurs. Aggradation is the vertical build-up of a sedimentary sequence. This usually occurs when there is

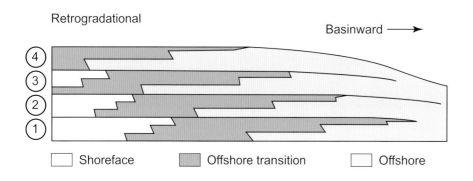

Figure 7.19 Retrogradational stacking of parasequences.

a relative rise in sea level produced by subsidence and/or eustatic sea-level rise, and the rate of sediment influx is sufficient to maintain the depositional surface at or near sea level. It occurs when sediment flux equals the rate of sea-level rise and produces aggradational stacking patterns in parasequences when the patterns of facies at the top of each parasequence are essentially the same (see Fig. 7.20).

An aggradational parasequence set is therefore one in which successively younger parasequences were deposited above one another, and there are no significant lateral shifts. The rate of creation of accommodation approximates the rate of accumulation and no deeper or shallower water facies will appear near the top or the base of the parasequence set. Aggradational stacking occurs in the highstand systems tract as the sea level falls from the maximum flooding level, gradually changing to progradational stacking as sea level falls. See Figure 7.21 showing main sequence boundaries.

To summarize, a progradational stacking pattern indicates a falling or regressive sea level, a retrogradational stacking pattern is indicative of a rising or transgressive sea level, and an aggradational pattern is evidence of a slowly rising sea level where sedimentation roughly equals sea-level rise. These interpretations can be applied to both carbonate rocks and clastic rocks. The geometries and stacking patterns of uncemented carbonates and clastics are similar. This is because both respond to changes in base level, both can be subdivided by similar surfaces, and both respond to wave and current movement similarly and may be transported.

There are, however, major differences in the sequence stratigraphy of these two sediment types. All clastics are transported to their depositional resting place, while carbonates are produced and accumulate *in situ*. Rates of carbonate production are linked to photosynthesis, so are dependent on depth,

with rates of formation greatest close to the air/sea interface. This means that carbonate facies and their fabrics are often used as indicators of sea-level position. An additional factor in carbonate accumulation is that they often have a biochemical and physico-chemical origin that is influenced by the chemistry of the water from which they are precipitated.

The main sequence boundaries of the Four Systems Tract Model are shown in the Wheeler diagram, Figure 7.21. There are illustrative vertical sections through the succession at A and B to show the positions of the main tract boundaries including sequence boundaries and maximum flooding surfaces.

Sequence stratigraphy allows geologists to analyse the stages in sedimentary basin infilling and to aid correlation of differing depositional environments across the basin. It is also a useful tool for assessing environmental changes through the duration of the basin development. Sequence stratigraphy has been an important adjunct to the integration of seismic and outcrop data.

The effect of lateral shifts in deposition such as described above is to create alternating layers of sandstones that are porous and permeable. Porosity is the amount of pore space in a rock and is expressed as the total volume of pore space over the total volume of the rock x 100. The porosity value is a measure of the amount of fluid, such as oil or water, that a rock might hold. The permeability of a rock is the ability of the rock to allow water or oil to pass through it, and can be expressed as a rate of flow. Those rocks that are impermeable do not permit the flow of water or oil. A highly porous and permeable rock is therefore capable of acting as a *reservoir* rock for hydrocarbons or water. Fine-grained rocks such as mudstones and shales are impermeable and act as hydrocarbon reservoir seals. Much of hydrocarbon exploration is a search for locations where suitable reservoir rocks are 'capped' or

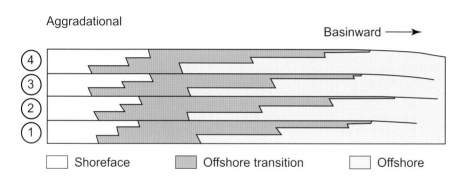

Figure 7.20 Aggradational stacking of parasequences.

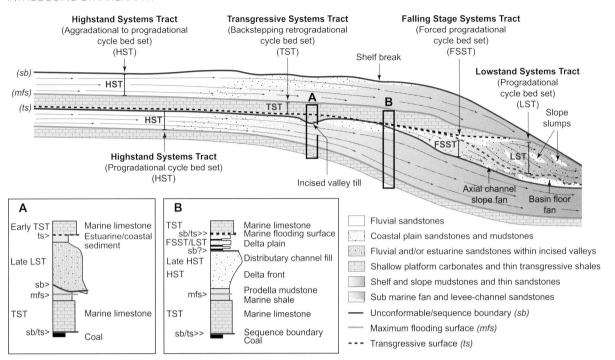

Figure 7.21 The main sequence boundaries of the Four Systems Tract Model for sequence stratigraphy, with vertical sections taken at A and B and the relative sea level shown for each tract.

sealed by mudstones or shales, and where hydrocarbons may have formed and migrated into these 'traps'. As one of the fundamental stratal units in sequence stratigraphy, the parasequence is of paramount importance in the recognition of sequences and the interpretation of the tectonic and depositional history of an area.

To illustrate parasequence cycles, it is convenient to use a *graphic log*. This is the standard method for collecting field data of sediments/sedimentary rocks in the field or from a drill hole. A graphic log gives an immediate visual impression of the section, and is a convenient way of making correlations and comparisons between equivalent sections from different areas. The logs consist of a description of the sequence of rocks, which are drawn to scale in tabular form with an appropriate symbol for each lithology, and including additional information on such features as grain size, sedimentary structures such as planar cross-bedding and trough cross-bedding, fossil content and colour. See Figure 7.22 as an example of a typical graphic log.

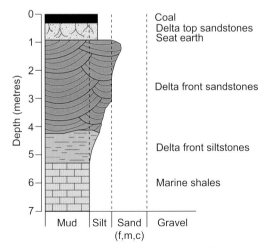

Figure 7.22 Typical graphic log giving information on rock type, succession details, grain size and structures such as planar and trough cross-bedding.

Cross-bedding is sedimentary layering within a stratum that is at an angle to the main bedding plane. These inclined surfaces are not the result of tilting or folding after deposition but are due to deposition by a flowing medium such as water or wind on bedforms such as ripples, dunes and delta slopes. See Figures 7.23 and 7.24.

Trough cross-beds have lower surfaces that are curved and truncate the underlying beds. Planar cross-bedding is formed mainly by the migration of large-scale, straight-crested ripples. A single unit of cross-bedded sediment is referred to as a set. Where a bed contains more than one set of the same type of structure, the stack of sets is referred to as a co-set (see Fig. 7.24).

Along with seismic data and other data sets, geologists use individual parasequence recognition and their vertical and lateral relationships and stacking geometries to interpret the depositional setting from the core provided by exploration well logs. Information from detailed analysis produces a sequence stratigraphic framework leading to accurate interpretations of depositional setting and predictions of lithofacies geometries in unknown portions of a basin. The following sections examine the ways in which parasequence and lithofacies identification is used in oil exploration to correlate separated successions.

Parasequence boundaries: interpreting parasequence deposition

A parasequence set approximately corresponds to a systems tract, and is categorized by its position within third order sequences (i.e., highstand, lowstand and transgressive). Stacking patterns of parasequences are used to define systems tracts. The parasequence is often characterized by a cycle of sediment that either coarsens or fines upwards so that the flooding surfaces are usually identified by abrupt and correlatable changes in the grain size of the sediments either side of that flooding surface. This change in grain size is often caused by the abrupt changes in energy that are associated with the waves or currents of the sea transgressing across the sediment interface. Such abrupt changes in grain size bounding a parasequence are recognizable in well logs, seismic sections and in outcrop, and are used to identify a parasequence cycle. Examples of these grain-size changes are best seen in the parasequences of coastal environments such as tidal flats, beaches, and deltas. Here parasequences correspond to individual prograding sediment bodies.

Figure 7.23 Cross-bedding in sandstones in the Canyons of the Escalante region, Utah, USA.

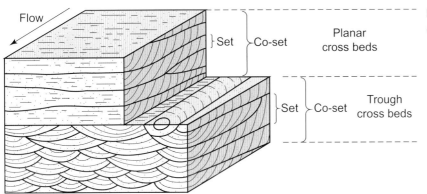

Figure 7.24 Planar cross-bedding and trough cross-bedding.

in Utah to Grand Junction in Colorado. A number of features make this an ideal area to study sequence stratigraphy:

- Strata are gently dipping, generally < 10° and there are no major folds or faults;
- The area is semi-desert with little vegetation to obscure the rocks;
- The cliff is cut by numerous canyons allowing a 3-D interpretation of the sections.

Although there are only negligible amounts of oil in the Book Cliffs, the area has served as an analogue for regions such as the North Sea, the Gulf of Mexico and the Niger Delta, where there are substantial hydrocarbon reserves. The concepts developed in sequence stratigraphy in Utah are thus readily transferred to other regions with similar sedimentary frameworks but more potential for oil and gas. Figure 7.29 shows the location of the Book Cliffs in Utah.

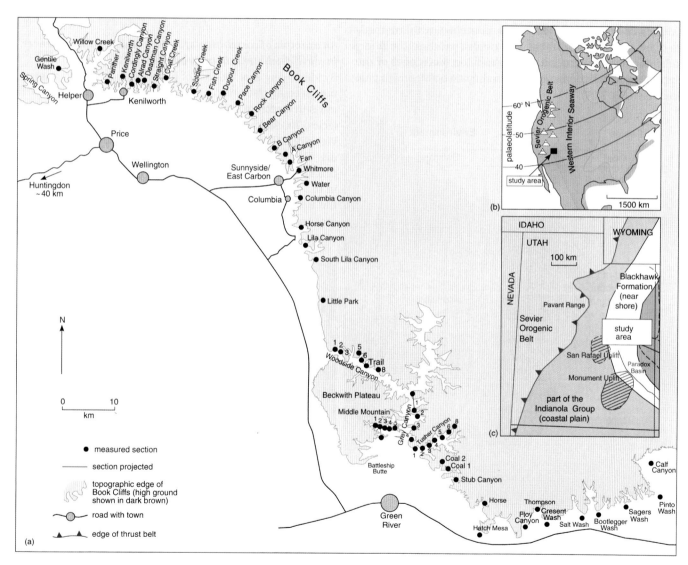

Figure 7.29 Showing the study area in the Cretaceous Blackhawk Formation, Utah, USA.

Measured sedimentary sections (Fig. 7.30A) in the Book Cliffs show high frequency parasequence sets with a range of settings from offshore shelf, lower shoreface, foreshore and upper shoreface and a coastal plain. A typical section through a wave-dominated shoreface passes from offshore siltstones into inter-bedded siltstones and hummocky cross-stratified sandstones of the offshore transition zone. Above these are the beds of the lower shoreface, in turn overlain by the upper shoreface and by the planar laminated sandstone beds of the foreshore. See graphic log, Figure 7.30B.

The field section illustrated in Figures 7.30 A and B shows the sedimentological setting of the parasequence sets within the succession. Using the criteria described earlier to recognize parasequence boundaries, it is possible to match sediment types and structures to one of the following depositional settings:

- foreshore and upper shoreface;
- sands with hummocky beds, burrows and current and wave ripples correspond to lower shoreface or delta;
- muds with burrows and planar beds correspond to offshore shelf.

(A)

Figure 7.30 **(A)** Bookcliffs high frequency clastic parasequence sets (after Coe et al, 2003). Lithofacies associations are: (a) Upper Shoreface; (b) Middle Shoreface; (c) Lower Shoreface; (d) Offshore Transition Zone; (e) Offshore. (Continued)

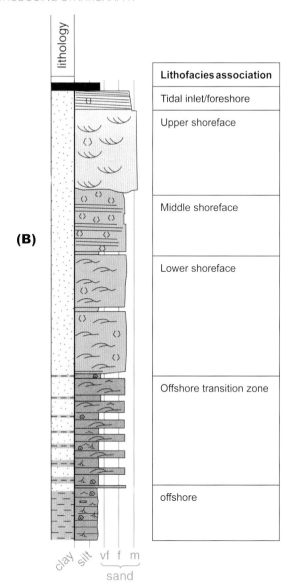

Lithofacies association
Tidal inlet/foreshore
Upper shoreface
Middle shoreface
Lower shoreface
Offshore transition zone
offshore

Figure 7.30 (Continued) **(B)** Graphic log of typical wave-dominated shoreface succession showing lithofacies association.

Consider the sedimentary succession in the Kennilworth section to the northeast of Helper in the Book Cliffs, Utah (see Fig. 7.29 for location). Following the same criteria this section can be divided into parasequences, with the top of the section

a parasequence boundary. The clastic section at Kennilworth can be divided into six parasequences showing the same lithofacies associations as Figure 7.30A and B:

- sands with trough cross-beds, burrows and current ripples and wave ripples corresponding to foreshore and upper shoreface;
- sands with hummocky beds, burrows and current and wave ripples corresponding to lower shoreface or delta front;
- muds with burrows and planar beds corresponding to shelf.

In Figure 7.31 each of the parasequences is marked with a triangle that broadens in the directions of coarser grain size, and each parasequence starts off with mudstones with burrows and is topped with sandstone. Only parasequence 6 contains the full range of lithofacies associations.

The next step in correlation is to use the same criteria to identify vertical sets of parasequences in the two additional measured sections from the Book Cliffs at Panther Canyon and Coal Canyon shown in Figure 7.32, and to use the information to correlate the three sections. Figure 7.32 shows graphic logs of the sedimentary successions at the three locations, Panther Canyon, Kennilworth and Coal Canyon (see Fig. 7.29 for location). As with the Kennilworth section, it is possible to match sedimentary types and structures to the following depositional settings:

- Coastal Plain;
- Foreshore and Upper Shoreface;
- Lower Shoreface or Delta Front;
- Shelf.

Note that an additional setting, the Coastal Plain, has been added compared to Figure 7.31. Using the recognition of differing depositional settings, coupled with abrupt changes in grain size or sedimentary structures, parasequence boundaries can again be identified in the three sections (see Fig. 7.33). Marine flooding surfaces in the three sections are used to separate the parasequences. Each parasequence in the three sections is a shallowing-upward cycle bounded by a marine flooding surface, so the lower surface of each cycle is the base of the deeper lithofacies (mudstone) layer that overlies the top of a shallowing-upward cycle. The upper boundary is the top of a shallower lithofacies (sandstone) layer that is overlain by a deeper lithofacies layer. The flooding surfaces bounding parasequences are not on the same

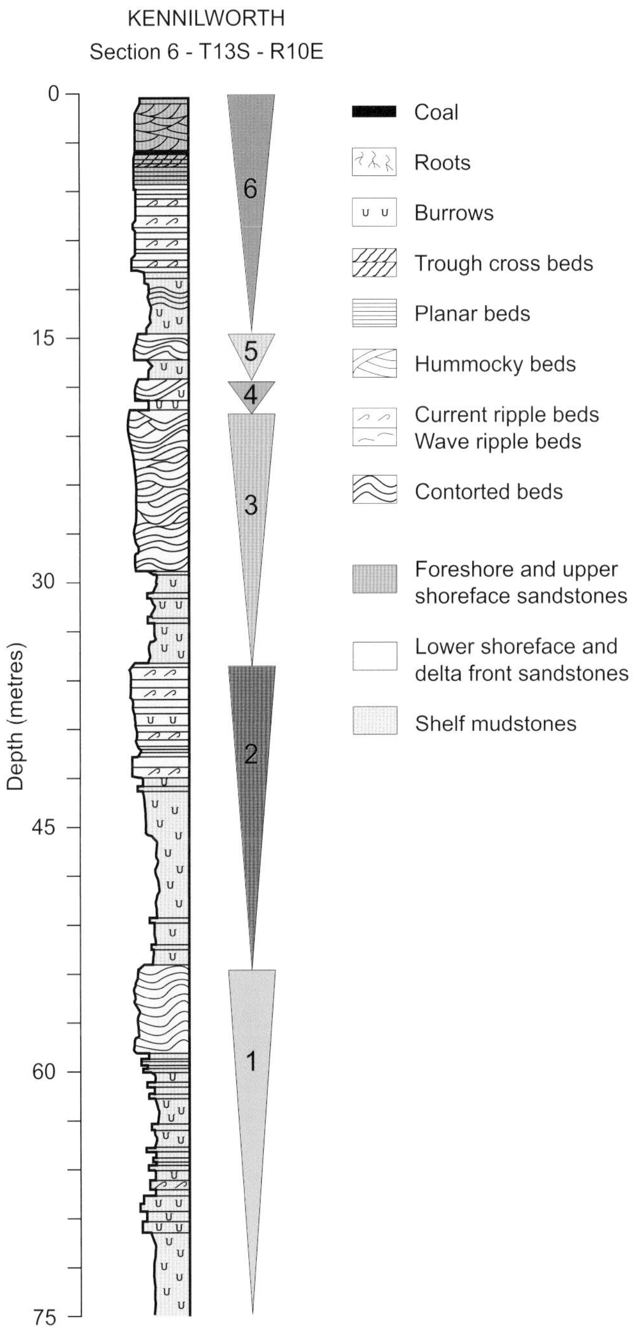

KENNILWORTH
Section 6 - T13S - R10E

Figure 7.31 Graphic log showing sequence stratigraphy of the Kennilworth Section.

scale as the regional transgressive surface that constitutes a sequence boundary.

Using the Kennilworth Section as a key, it is possible to correlate it with the sections at Coal Canyon to the east and Panther Canyon to the west by recognizing the similar depositional settings:

1 Coal, sand trough-cross beds and burrows match Coastal Plain;
2 Sands with trough cross-beds, burrows and current and wave ripples match Foreshore and Upper Shoreface;
3 Sands with hummocky beds and burrows and current and wave ripples match Lower Shoreface or Delta Front;
4 Muds with burrows and planar beds match Shelf.

This example shows that by integrating seismic, borehole and outcrop data, sequence stratigraphy can be used to study sedimentary rock relationships using a time-stratigraphic framework of cyclical, genetically related contemporaneous strata bounded by surfaces of erosion or non-deposition. Within sequence stratigraphy the parasequence is the fundamental strata unit, and is of paramount importance in recognizing sequences and interpreting the depositional and tectonic history of an area from its sequence development. In the search for hydrocarbons it provides an effective and systematic approach to the recognition of potential reservoir and cap-rocks.

Biostratigraphy
Microfossils in biostratigraphy
The basis of biostratigraphy as discussed in chapter 2 is the Law of Faunal Succession, where fossil plants and animals succeed each other in a definite order. **Microfossils** are small fossils that can only be studied and identified using a microscope. They may be either a fragment of a larger animal or an entire minute organism, and they play an extremely important role in biostratigraphy. In many ways microfossils are ideal index fossils, since they are and were abundant and widely distributed across a range of facies and environments. In addition, they evolved rapidly and so appeared and disappeared from the stratigraphic record, and like the graptolites discussed earlier, they can be used to define narrow time zones. They also record palaeo-environmental variations that can be used in sequence stratigraphy, as detailed earlier in this chapter.

A further role for microfossils, discussed earlier in this chapter when dealing with the use of foraminifera in

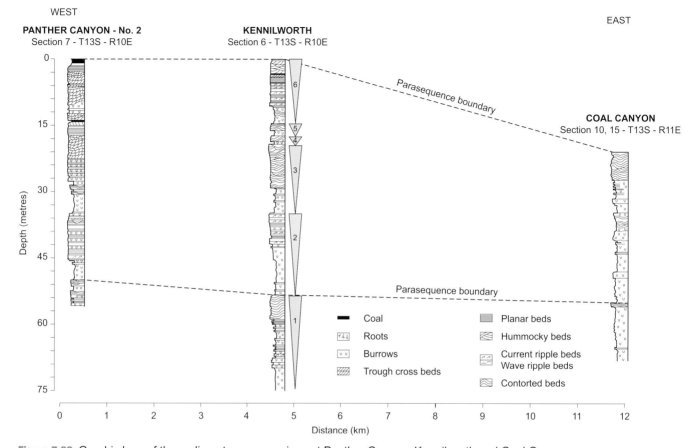

Figure 7.32 Graphic logs of the sedimentary successions at Panther Canyon, Kennilworth and Coal Canyon.

chemostratigraphy, is their capacity to record changes such as the oxygen isotope character of their environment. These variations, along with changes in assemblages, are fundamentally important in the investigation of climate change throughout geological history. Some varieties of microfossils can be used as palaeo-thermometers; that is, with post-burial heating they undergo irreversible colour changes that are an indication of hydrocarbon maturity levels.

Microfossils in petroleum exploration

It is a fact in petroleum exploration that most of the rock samples recovered by drilling are in the form of small cuttings of rock produced by the action of the diamond drill bit and brought to the surface in the fluid, 'mud', used to lubricate and cool the drill head. Unlike the larger macrofossils such as trilobites or vertebrates used in conventional biostratigraphy, which would be rendered unrecognizable by being broken up, microfossils can be recovered complete and undamaged.

Microfossils used in the exploration for hydrocarbons are divided into four main groups based on the composition of their shells or hard parts. See table 7.3 showing the principal microfossil groups.

Foraminifera

Foraminifera are single-celled organisms with a shell or test. They occur as abundant fossils throughout most of the

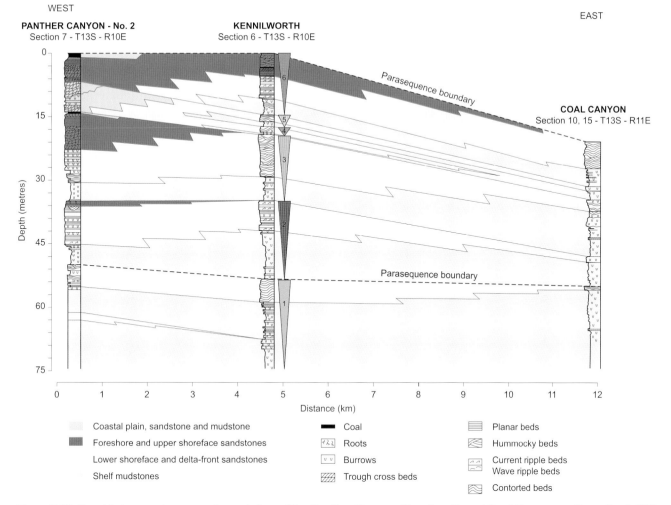

Figure 7.33 Graphic logs and proposed correlation of the Panther Canyon, Kennilworth and Coal Canyon sections, Book Cliff outcrops, Utah, USA.

Table 7.3 The principal groups of microfossils used in biostratigraphy.

Composition	Fossil Group
Calcareous	Foraminifera, ostracods, calcareous nannofossils
Phosphatic	Conodonts
Siliceous	Radiolarians, diatoms, silicoflagellates
Organic walled	Chitinozoans, pollen and spores, acritarchs, dinoflagellates (collectively palynomorphs)

Phanerozoic and are predominantly calcareous in composition. Currently they occupy a wide range of environments in the world's oceans, many species living on or in the sea-bed sediments, with a number of species planktonic. Foraminifera are sufficiently abundant to form an important sediment source in some beach environments, while in some deep-ocean environments planktonic foraminifera shells are almost the sole sediment source. They first occur in the Cambrian and have evolved continuously since then,

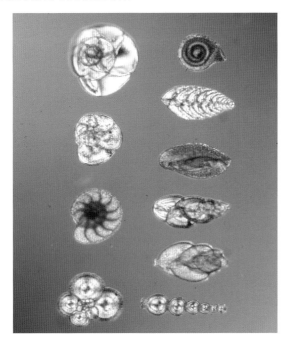

Figure 7.34 Foraminifera.
Dr. Norbert Lange/Shutterstock

producing a large number of species that are useful as zone fossils (see Fig. 7.34).

Conodonts

Conodonts are an extinct group of animals resembling eels. For many years they were only known as tooth-like microfossils of phosphate composition (calcium carbonate fluorapatite, also found in human bones), found in marine deposits, commonly in black shales and associated with graptolites. They range in size from 200 micrometres to 5 millimetres. The term 'conodont', from the Greek meaning *cone* and *tooth*, was first used in Russia in the 1850s and by the 1920s their usefulness in biostratigraphy had been recognized. Although information about the soft parts of the animal remains limited, it is thought now to be an eel-like creature, with a number of conodont elements forming a more complex feeding apparatus similar to jawless fishes such as hagfish (see Fig. 7.35).

Morphologically there is a range of conodont shapes, from single cones, blades and bars to platform types. The earliest conodonts are found in Precambrian rocks in Siberia, becoming more abundant in the Cambrian, with increased diversity

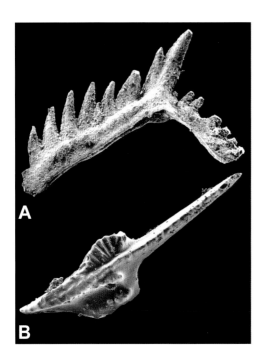

Figure 7.35 Conodonts from the Glen Dean formation (Chester) of the Illinois basin.

in the Ordovician and through to the Devonian. They remained abundant and widespread in the Carboniferous but declined from the late Carboniferous onwards, becoming extinct in the late Triassic. This wide-ranging distribution in the Palaeozoic when other microfossil groups were either non-existent or sparsely present means that they have become the main microfossils for palaeontologists working on sediments from the Ordovician to the Triassic periods. Conodont elements are among those microfossil groups that are useful palaeo-thermometers as a proxy for thermal alteration in the host rock. The phosphate in the conodont element undergoes pre-dictable colour changes, which are then measured against a conodont alteration index.

Radiolaria

The **radiolaria** are protozoa with diameters 0.1–0.2mm with complex skeletons usually composed of **silica**. They are found throughout the ocean environment, and the siliceous ooze that covers extensive areas of the ocean floor is composed of these skeletal remains. As with other groups of microfossils, they evolved rapidly and are important zone fossils from the Cambrian onwards (see Fig. 7.36).

Radiolaria have an unusually long range, from late Precambrian to Recent. Since they have a siliceous skeleton they are used to date sediments that do not include calcareous fossils, either because the rocks are too old or because deposition was below the **Carbonate Compensation Depth (CCD)**. That is the depth in the oceans at which the rate of carbonate accumulation equals the rate at which it is dissolving, so that below this level no calcite is preserved. Rocks known as **cherts** are fine-grained sedimentary rocks composed of micro-crystalline quartz often derived from radiolarian skeletons. Marine geological settings such as **ophiolites** or **accretionary terranes,** which result from plate collisions, have been successfully investigated using radiolaria in the absence of any other zone fossils.

Diatoms

Diatoms are a major group of photosynthesizing algae that inhabit almost every aquatic environment. They exist within a cell wall known as a frustule, which is siliceous in composition. Frustules show a wide range of forms but are mostly bilaterally symmetrical, hence the name diatom (see Fig. 7.37). They are commonly between 20 and 200 microns

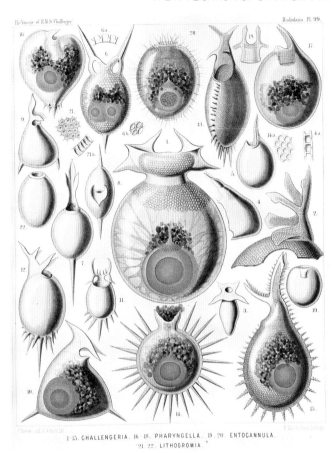

Figure 7.36 Radiolaria illustration from the Challenger Expedition, 1873-76.

in diameter or length, and can be both solitary and colonial. They often occur in sufficient numbers to form sediment deposits known as **diatomite** or diatomaceous earth. These deposits have a wide range of uses including filtration processes, as insecticides and as filler in toothpastes. Historically, Alfred Nobel discovered that the explosive nitro-glycerine was more stable when mixed with diatomite. This mixture he patented as dynamite.

Diatoms first appeared in the fossil record in the Jurassic, becoming more prominent from middle Cretaceous onwards. The evolutionary history of diatoms has been marked by several major changes in floral diversity (known as floristic

122

Figure 7.37 Marine diatoms: Diatoms Suriella magnified.

turnovers) and these have been used in biostratigraphic correlations. For example, the extensive deep-sea drilling programmes of recent years have allowed the construction of a biostratigraphy based on diatoms for most of the Cenozoic. Diatom communities are also useful indicators of environmental conditions past and current. They are extensively used in water quality studies.

Silicoflagellates are a small group of unicellular algae found in marine environments which, like radiolaria, produce a siliceous skeleton. They are recognized as microfossils from the early Cretaceous and have been used in biostratigraphical correlations in a number of marine environments, often in association with diatoms.

Palynomorphs

Palynomorphs are microfossils, both plant and animal structures, 5–500 microns in diameter and composed of organic material, usually compounds resistant to decay such as sporopollenin or related compounds such as **chitin**. Sporopollenin is a chemically inert biological polymer that forms the outer walls of plant spores and pollen grains. Chitin is the main component of the outer covering of insects, and also occurs in the cell walls of certain fungi and algae. The chemical characteristics of these compounds mean they have a high potential for preservation in sediments.

Palynology is the study of fossil and living pollen, spores and other similar palynomorphs. These range from the Precambrian to Recent. Palynology is a branch of micropalaeontology that makes an important contribution to biostratigraphy.

Spores and pollens

The most important and widespread group of palynomorphs are the spores and pollen derived from terrestrial plants. The first land plants appeared towards the end of the Silurian with the development of the first land ferns, and the first pollens are found in the Carboniferous with the first gymnosperms. These have seeds that are 'naked' (*Greek gumnos, naked*), that is the seeds are unprotected by a fruit, for example conifers. The evolution of flowering plants, the **angiosperms,** which have seeds enclosed within a **carpel**, meant greater diversity in pollen forms. The angiosperms include shrubs, grasses and most trees.

Since spores and pollen grains are so readily transportable by water and wind, they are found in almost all environments and in a wide range of sediments. Occurring from the Palaeozoic to Recent, they are important in biostratigraphy, particularly in the study of terrestrial deposits where other macro- or micro-fossil groups may be absent or scarce.

Acritarchs

The term **acritarchs** includes any of a group of single-celled marine planktonic microfossils of uncertain origin and affinities. Their outer wall is organic and it is thought they have affinities to algae. They are the predominant component of organic-walled microfossils in the Palaeozoic and so have a role in the study in biostratigraphical investigations such as delineating the base of the Cambrian.

Chitinozoa

Chitinozoa are members of a group of flask-shaped marine microfossils of uncertain derivation, characterized by an organic wall (see Fig. 7.38). They are important in Palaeozoic biostratigraphy, particularly in the Ordovician to Devonian periods, due to their widespread distribution in a wide range of marine environments coupled with rapid rates of evolution.

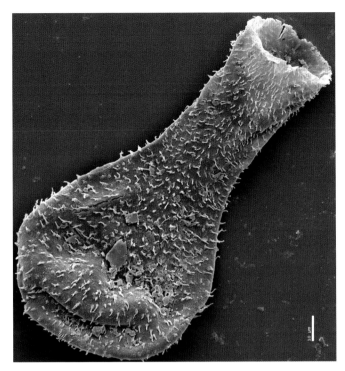

Figure 7.38 Scanning electron micrograph of a late Silurian chitinozoan from the Burgsvik Beds in Sweden showing flask shape. The scale bar is 10 micron.

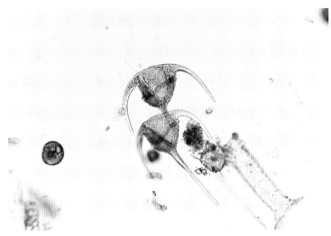

Figure 7.39 Ceratium hirundinella, Dinoflagellata, Trias -Recent.

Dinoflagellates

Dinoflagellates are an organic-walled group of microfossils that are predominantly found in marine environments, but are also found in fresh water (see Fig. 7.39). A bloom of certain modern dinoflagellates can produce a coloration of the sea in what is known as a 'red tide', which is toxic to humans if consumed via contaminated shellfish. They are important in post-Triassic biostratigraphy, particularly in non- or low-carbonate rocks such as siltstones and mudstones. They are especially useful in sequence stratigraphy, where their species diversity can be used in the identification of the sequence tract, as well as for palaeoclimate and palaeo-environment reconstructions.

Summary

The use of biostratigraphy in the hydrocarbon industry is well established, both as a dating tool and in the investigation of the environment of deposition of the sediments. The precise role of micropalaeontology in petroleum exploration is the determination of the age and stratigraphic position of the sequence, either in outcrop or via data from a drill hole. In addition, microfossils have an important role in the building of palaeoclimate models, the measurement of palaeo-temperatures, and other environmental indicators such as sea-level rises. The combination of biostratigraphy and seismic and geomagnetic profiles as the basis of sequence stratigraphy has allowed a connection to be made between the seismic data and absolute ages. This has proved a powerful tool in regional correlation. In exploration for hydrocarbons, microfossils such as conodonts and foraminifera can be used as proxy environmental indicators. Conodonts are used as palaeo-thermometers in estimating thermal maturity of the host rock in petroleum exploration, and foraminifera are proxies for ocean temperatures, using the oxygen isotope ratios of their calcareous shells.

Epilogue

The key to understanding the science of geology is an appreciation of geological time. The subdivision of geological time, and therefore Earth history, is through the principles of stratigraphy. Essentially this involves placing rock strata in the order of their deposition (*relative time*) and calibrating that sequence (*absolute time*) by the techniques of radiometric dating. Time can then be characterized as a succession of historical events, placed in the order of their occurrence and with a measure of their duration. The recognition of Time's Arrow and Time's Cycle in stratigraphy attempts to reconcile the apparent conflict between Uniformitarianism and Catastrophism. The contrast is between Earth history as a sequence of linked events moving in a single direction (time's arrow) and the alternative view that geological phenomena have no significance as individual occurrences, but instead represent periodicity (time's cycle). In his book *The Nature of the Stratigraphical Record*, Derek Ager coins the term *Catastrophic Uniformitarianism* to explain the co-existence of the two concepts, while arguing for a greater role for catastrophic events than has hitherto been acknowledged.

The major subdivisions of the stratigraphic column result from factors such as changes in tectonic plate configuration, the evolution of the atmosphere, periods of glaciations and sea-level changes, both relative and eustatic. Superimposed on these terrestrial processes are extra-terrestial influences such as meteorite or comet impacts. All of these phenomena have driven the development and evolution of life on Earth. The fossil record of flora and fauna has been the basis of the structure of the stratigraphic column with the Proterozoic Eon (*earlier life*) and the Phanerozoic Eon (*visible life*) accounting for around 2.5byr of Earth history and reflecting this evolutionary progression.

This book has attempted to explain the concept of *deep time* and the ways it can be measured and subdivided according to Steno's principles of stratigraphy and the techniques of radiometric dating. It has charted the development of stratigraphy as a fundamental part of the science of geology and has described the ways in which stratigraphy in its various forms is used in the search for resources, for a world seemingly in need of an ever-increasing supply of commodities. In a climate of opinion where attitudes in the developed world towards the use of fossil fuels appear to be under-going a profound change, there is a case to be made for an enhanced understanding of stratigraphic principles. The world will not be weaned off fossil fuels overnight, and there is also greater demand for the strategic metals required for modern technological applications. These metals are often hosted within sedimentary rocks. The more information there is about the existence and extent of such reserves, the better will be their management and use in a controlled way for the benefit of all on the planet.

Glossary

A

absolute age, – time |7|: the dating of geological events in years.

accretionary terranes |121|: a fault-bounded geological entity characterized by a distinctive stratigraphic or structural history, which differs markedly from the surrounding areas. A terrane is a fragment of crust broken from one tectonic plate and accreted or sutured to another plate.

acritarchs |122|: a group of microfossils that range from Precambrian times to present that show affinities with algae and egg cases.

amniote egg |69|: an egg that is waterproof and so can be laid on land.

angiosperms |122|: flowering plants.

angular unconformity |34|: an unconformity where horizontally parallel strata are deposited on tilted and eroded layers, producing an angular discordance with the overlying horizontal layers.

anthropocene |51|: the proposed most recent geological time period based on global evidence that Earth system processes are now being altered by human activity.

Archaean |49|: an eon, a unit of geological time spanning the period 4000–2500mya.

asthenosphere |25|: weak layer beneath the *lithosphere*, capable of solid-state flow. Over which the tectonic plates slide.

B

Banded Iron Formations (BIFs) |60|: a distinctive sedimentary rock mostly confined to the Precambrian. Usually consists of repeating thin layers of silver to black iron oxides, magnetite or haematite, alternating with bands of iron-poor shales and cherts, often red in colour.

basalt |3|: a dark, fine-grained extrusive igneous rock.

basement rocks |47|: are those below a sedimentary cover and which are igneous or metamorphic in origin.

BCE |1|: Before the Common Era. Replacement for BC (Before Christ), with the Common Era the equivalent to AD (Anno Domini).

bed |85|: a distinct horizontal unit of sedimentary rock, representing a single episode of sediment deposition.

biostratigraphy |7|: the part of stratigraphy that uses fossils and fossil assemblages to date rock units.

biozone |11|: a short time unit identified by a distinctive fossil or group of fossils.

borehole logging |91|: the practice of making a detailed record, a well log, of the geological formations penetrated by a borehole. The practice is also known as well logging.

boulder clay |52|: ground moraine produced by the grinding action at the base of a moving ice sheet. Consists of a stiff clay containing rock fragments of all sizes.

C

Cambrian explosion |61|: the event around 540mya during the Cambrian Period when complex animals with mineralized skeletons appeared relatively suddenly in the fossil record.

Cambrian |17|: a division of geological time in the Palaeozoic Era between 542 and 488mya.

cap carbonate |87|: layers of distinctly textured carbonate rocks that typically overlie Neoproterozoic glacial deposits.

Carbonate Compensation Depth (CCD) |121|: the depth in the oceans below which the rate of accumulation of calcite (calcium carbonate) equals the rate of carbonate dissolution, so that no calcite is preserved.

Carboniferous System |16|: a term now including the two geological periods Mississippian and Pennsylvanian, 359–299mya.

carpel |122|: the female reproductive organ of a flower, consisting of an ovary, a stigma and a style.

Catastrophism |23|: the theory, advanced by Georges Cuvier, that changes in the Earth's crust resulted chiefly from sudden, violent and unusual events.

Cenozoic |20|: an era, a division of geological time between 65mya and the present day.

chemostratigraphy |7|: the study of chemical variations within sedimentary rocks to determine their stratigraphic relationships.

chert |60|: a fine-grained sedimentary rock composed of microcrystalline silica.

chitin |122|: a hard material made of protein found in the carapaces of arthropods and in certain microfossils.

chitinozoa |122|: marine microfossils of uncertain derivation, characterized by an organic wall.

chronostratigraphy |7|: that part of stratigraphy which is concerned with absolute ages.

clastic |102|: sedimentary rock formed from particles (clasts) that were mechanically transported.

constructive (divergent) plate margins |25|: where new plate is created by divergent movement of two adjacent plates, marked by a mid-ocean ridge or continental rift.

coccoliths |74|: minute circular plates composed of calcite, the main component of Cretaceous chalk.

conodonts |120|: phosphatic tooth-like elements from an extinct group of eel-like animals. Useful for biostratigraphic dating in the Palaeozoic.

correlation |4|: matching rock units of the same time across geographical distances.

craton |52|: the stable part of a continental interior, unaffected by contemporary orogenic activity.

Cretaceous |16|: a division of geological time in the Mesozoic Era between 145 and 65mya.

crust [13]: the uppermost layer of the Earth. *Continental crust* is that part of the crust underlying a continent. *Oceanic crust* is composed mainly of basalt and formed at constructive plate margins.

D

daughter atoms [43]: produced by or resulting from the decay of a radioactive element.

decay series [43]: the sequence of elements formed when a radioactive element such as thorium or uranium decays until the next stable atom, usually lead, is produced.

deep time [24]: encapsulates the idea that geological time is vast and measured in millions and billions of years.

delta [11]: a landform resulting from deposition of sediment carried by a river as the flow leaves its mouth and enters slower-moving water such as in a sea or lake.

destructive (convergent) plate margins [25]: plate boundary characterized by the convergent movement of adjacent plates and by the destruction of oceanic crust or the collision of continents.

Devonian [18]: a division of geological time in the Palaeozoic Era between 416 and 359mya.

diagenesis [44]: the changes that a sediment undergoes as it lithifies to become a sedimentary rock. These will include dewatering, compaction and cementation.

diatomite [121]: silica-rich sediment composed almost entirely of diatoms.

diatoms [121]: small microscopic algae.

dinoflagellates [123]: an organic-walled group of microfossils predominantly found in marine environments.

disintegration theory [42]: general theory of radiation put forward by Ernest Rutherford, which stated that the atoms of radioactive bodies are unstable and every second a certain fixed proportion of them break up with the release of energy in the form of alpha-particles.

E

Ediacaran Period [48]: a division of geological time in the Upper Proterozoic of the Precambrian, from 635–542mya.

Eocene [21]: a subdivision of the Palaeogene Period 55.8–33.9mya.

eon [20]: a major division of geological time, such as the Phanerozoic, that contains the smaller units, eras.

epoch [85]: a subdivision of geological time within a geological period.

era [20]: a division of geological time, such as the Mesozoic, that consists of a number of geological periods.

Eukaryota [62]: organisms formed of cells with a distinct nucleus surrounded by a nuclear membrane.

eustatic sea-level changes [57]: when the sea level changes due to an alteration in the volume of water in the oceans or a change in the shape of the ocean basin.

evaporite [72]: chemogenic sediments precipitated from water following the concentration of dissolved salts by evaporation. Comprise principally the minerals gypsum, halite and anhydrite.

extrusive [45]: igneous rocks are those resulting from volcanic activity at the Earth's surface.

F

facies [35]: a group of sedimentary rocks and structures indicative of a given depositional environment.

Foraminifera [97, 118]: microscopic unicellular animals with ornate shells that live in a variety of marine and brackish-water environments. They are important for biostratigraphic and palaeoclimate studies.

formation [11, 85]: in stratigraphy the primary unit consisting of a succession of rocks useful for mapping or description.

G

Geomagnetic Polarity Time Scale (GPTS) [95]: in geomagnetism a record of the many episodes of the reversal of the Earth's magnetic field.

Global Boundary Stratotype Section and Points (GSSP) [85]: internationally agreed reference points on a stratigraphic section used to define the lower boundary of a stage on the geological timescale.

Global Standard Stratigraphic Ages (GSSA) [86]: in stratigraphy a chronological reference point used to define the boundaries between different geological periods, epochs or ages on the overall geological timescale.

gneiss [3]: a coarse-grained regional metamorphic rock that shows compositional banding of minerals.

granite [3]: a pale, coarse-grained intrusive igneous rock.

graptolite [12]: a class of colonial organisms with skeletons of organic material, which were important zone fossils in the Palaeozoic.

greenhouse gas [52]: a gas that contributes to the warming of the atmosphere (the Greenhouse Effect) by absorbing infrared radiation. Carbon dioxide is an example.

greywacke [17]: a poorly sorted sandstone containing more than 15% of fine-grained matrix.

group [85]: in stratigraphy a lithostratigraphic unit, part of the geological record, consisting of two or more formations.

gymnosperm [71]: a group of seed-bearing plants including conifers, dominant in the Mesozoic.

H

Hadean [49]: an informal name for the division of time between the formation of the Earth 4567mya and 4000mya.

half-life [43]: in radiometric dating the time taken for half the initial number of parent atoms to decay to daughter atoms.

Holocene [22]: in stratigraphy relating to or denoting the present epoch. It is the second epoch in the Quaternary Period and followed the Pleistocene.

I

igneous rocks [3]: rocks formed by the crystallization of molten magma.

index fossil [11]: in stratigraphy a fossil that is useful for dating and correlating the rock in which it is found.

International Chronostratigraphic Chart [83]: a hierarchy of chronostratigraphic units (Systems, Series and Stages) are based.

International Commission on Stratigraphy [83]: a constituent body of the International Union of Geological Sciences with the primary objective of precisely defining the global units of the International Chronostratigraphic Chart.

intrusive [45]: a body of molten igneous rock that has intruded or invaded older rocks beneath the surface, and subsequently solidified.

isostatic readjustment [35]: the process of gravitational equilibrium at the Earth's surface, in which topographic variations are balanced by density variations beneath.

isotope [42]: one of several forms of one element, all having the same number of protons in the nucleus but with differing numbers of neutrons, and so different atomic weights.

J

Jurassic System [16]: a division of geological time in the Mesozoic Era between 200 and 145mya.

K

kerogen [73]: naturally occurring solid, insoluble, organic material in sedimentary rocks that can yield petroleum products on heating.

L

limestone [3]: a sedimentary rock composed mainly of the calcium carbonate mineral, calcite.

lithified [3]: a state reached after the processes that convert a newly deposited sediment into an indurated rock.

lithosphere [25]: the strong upper layer of the Earth, including the crust and the upper part of the mantle.

lithostratigraphy [7]: the definition of units of rocks in terms of their lithology and their correlation from area to area.

M

magma [3]: molten rock material that forms igneous rocks on cooling.

magnetostratigraphy [7, 95]: the definition of units of rocks in terms of their magnetic characteristics.

mantle [19]: the central layer in the Earth between the crust and the core (between 50 and 2900km).

marble [3]: the metamorphic equivalent of limestone.

marine flooding surface [105]: a surface separating younger from older strata, across which there is evidence of an abrupt increase in water depth. It may be accompanied by minor submarine erosion or non-deposition, with a minor hiatus indicated. This surface often forms the maximum flooding surface (mfs), marking the boundary between the prograding highstand systems tract and the top of the transgressive systems tract.

mass extinction events [23]: the extinction of a large number of species within a relatively short period of geological time, often due to catastrophic global events or rapid widespread environmental change.

maturation [74]: process of a source rock becoming capable of generating oil or gas when exposed to appropriate temperatures and pressures.

maximum flooding surface [106]: a surface of deposition at the time when the shoreline is at its maximum landward position, i.e. the time of maximum transgression. The surface marks the time of maximum flooding or transgression of the shelf, and it separates the transgressive and highstand systems tracts.

member [85]: in stratigraphy a division of a formation, generally of distinct character or only local extent.

Mesozoic [20]: the era of geological time between the Palaeozoic and the Cenozoic, occurring between 251 and 65mya.

metamorphic rocks [3]: those rocks changed by the effects of heat or pressure or both, with the development of new minerals and structures.

microfossil [117]: a fossil so small it can only be seen with the aid of a microscope.

Miocene [21]: a subdivision (epoch) of the Neogene period 23 to 5.3mya.

N

Neogene [21]: a division of geological time in the Cenozoic Era between 23 and 2.6mya.

O

Oligocene [21]: an epoch of the Palaeogene Period extending from 33.9 to 23mya.

ophiolite [121]: a sequence of rock types interpreted as pieces of oceanic crust.

Ordovician [19]: a division of geological time in the Palaeozoic Era between 488 and 444mya.

orogeny [14]: process in which a section of the Earth's crust is folded and deformed by lateral compression to form a mountain range.

P

Palaeocene [21]: the first epoch of the Palaeogene Period in the modern Cenozoic era.

Palaeogene [21]: a division of geological time in the Cenozoic Era between 65 and 23mya.

Palaeomagnetism [95]: the study of the magnetism in rocks induced by the Earth's magnetic field at the time of their formation.

Palaeozoic [20]: the era of geological time between 542 and 251mya.

palynology [122]: the study of fossil and living pollen, spores and other similar palynomorphs.

palynomorphs [122]: microfossils, both plant and animal structures, composed of organic material, usually compounds resistant to decay such as chitin.

Pangaea [29]: a supercontinent composed of Gondwana to the south and Laurasia to the north, which existed from the Permian until its breakup during the Cretaceous.

parasequence stacking pattern [107]: aggradational parasequence stacking occurs when successively younger parasequences are deposited above one another and there are no significant lateral shifts. Progradational parasequence stacking occurs when successively younger parasequences are deposited further basinwards. Retrogradational parasequence stacking occurs when successively younger parasequences are deposited further landwards.

parent atoms [13]: refers to the atom that undergoes radioactive decay in a nuclear reaction.

period [16]: a division of geological time such as Cambrian or Ordovician that contains epochs.

Permian [20]: a division of geological time in the Palaeozoic Era between 299 and 251mya.

Phanerozoic Eon [20]: a major division of geological time between 542mya and the present day. Phanerozoic means 'visible life'.

plane of unconformity [14]: a surface of erosion or non-deposition separating younger rocks from older.

planktonic [12]: living in the surface waters of the oceans.

plate [25]: a relatively stable piece of the lithosphere that moves independently of adjoining plates.

plate tectonics [1]: study of the formation, movement and interaction of lithospheric plates on the Earth's surface.

Pleistocene [22]: a subdivision (epoch) of the Quaternary period between 2.5mya and 12,000 years ago.

Pliocene [21]: a subdivision (epoch) of the Neogene period 5.3 to 2.5mya.

Proterozoic [20]: an Eon, a major division of geological time between 2500 and 542mya.

Q

Quaternary [21]: unit of geological time in the Cenozoic Era, 2.5mya to Recent.

R

Radiolaria [121]: microscopic animals with shells composed of silica that form part of the planktonic life in the upper part of the ocean.

radiometric dating [60]: method of obtaining ages of geological materials by measuring the relative abundances of a radioactive element and the lighter element produced from it by radioactive decay.

regression [35]: periods when the sea level of the oceans falls and the continental shelf area may be exposed.

relative age [7]: the dating of geological events according to the system of successive eras, periods and epochs.

remanent magnetization [94]: permanent magnetization acquired by igneous rocks as they cool below the Curie Point in the presence of the Earth's magnetic field.

rock cycle [3]: the succession of events that results in the transformation of Earth materials from one rock type to another.

Rodinia [29]: a Late Proterozoic supercontinent that assembled around 1.3bya and broke up around 750mya.

S

sandstone [3]: a sedimentary rock composed of consolidated and cemented sand grains.

sea-floor spreading [25]: the theory of the generation of ocean crust at ocean ridges and its destruction at ocean trenches.

sedimentary rocks [3]: rocks that form either by the cementing of grains broken off pre-existing rocks, or by the precipitation of mineral crystals out of water near the Earth's surface.

seismology [97]: the study of vibrational seismic waves, earthquakes and the interior structure of the Earth.

sequence boundary(ies) [106]: significant erosional unconformities and their correlative conformities, resulting from a fall in sea level and erosion of the subaerially exposed sediment surface of the earlier sequence or sequences.

sequence stratigraphy [7, 15, 101]: that branch of stratigraphy that seeks to subdivide sedimentary deposits into units bounded by unconformities.

series [84]: in stratigraphy, consists of the rocks formed during an epoch.

silica [121]: an oxide of silicon that is found in a variety of forms, commonly as the mineral quartz, an important component of sedimentary rocks. Used by some organisms such as radiolarians and diatoms to form skeletal components.

silicoflagellates [122]: belong to a small group of marine planktonic organisms with siliceous skeletons. They range from early Cretaceous to present and are used in biostratigraphy in environments where calcareous microfossils are scarce or would be dissolved.

Silurian [17]: a division of geological time in the Palaeozoic Era between 444 and 416mya.

Snowball Earth [52]: the hypothesis that the Earth's surface was entirely frozen at least once in the Precambrian.

stage [85]: a small unit of rock contained within a short subdivision of geological time.

Stone Age [85]: a broad prehistoric period during which stone was widely used to make implements with an edge, a point or a percussion surface. Came to a gradual end with the advent of metalworking, with the use of bronze, then iron.

strata [3]: layers in sedimentary rocks.

stratigraphy [1]: that branch of geology concerned with the ordering of the rock succession and its dating.

stratigraphic column [7]: a representation used in geology to describe the vertical location of rock units in a particular area. The oldest rocks are at the bottom and the youngest at the top.

striation [54]: scratches left on bedrock by fragments of rock embedded in overriding ice, often showing the direction of ice movement.

subduction [25]: the process whereby an oceanic plate descends into the mantle along a subduction zone.

supercontinent [29]: a large continental mass consisting of several components that previously, or subsequently, were continents themselves.

system [16]: a major division of the geological rock succession equivalent to the geological period used for the division of that time; for example, the Cambrian System covers the rocks of the Cambrian Period.

system tracts [106]: in sequence stratigraphy, suites of co-existing depositional systems such as coastal plains, continental shelves and submarine fans.

T

Tertiary [16]: a now obsolete name of a geological period in the Cenozoic Era, replaced by Palaeogene and Neogene.

till [52]: unsorted glacial sediment derived from the erosion and entrainment of material by moving ice. The sediment is then deposited to form various types of moraines.

transgression [35]: a geological event during which sea level rises relative to the land and the shoreline moves toward higher ground, resulting in flooding.

Triassic [19]: a division of geological time in the Mesozoic Era between 251 and 200mya.

U

unconformity [14]: a buried erosional or non-depositional surface separating rock strata of differing ages, indicating that sediment deposition was not continuous. See *plane of unconformity*.

Uniformitarianism [14]: the theory that Earth's geological processes throughout geological history result from the action of continuous and uniform processes, acting with essentially the same intensity in the past as they do in the present.

V

vascular plants [69]: plants characterized by having conducting tissue for water and minerals.

Z

zone fossils [11]: fossils that define a narrow time period known as a zone.

Further reading

Ager, D.V. (1981) *The Nature of the Stratigraphic Record*. 2nd edn. London: The Macmillan Press.

Coe, Angela L. (ed.) (2002) *The Sedimentary Record of Sea-level Change*. Cambridge: Cambridge University Press.

Jones, S.J. (2015) *Introducing Sedimentology*. Edinburgh: Dunedin Academic Press.

Lyle, P. (2016) *The Abyss of Time: a Study in Geological Time and Earth history*. Edinburgh: Dunedin Academic Press.

McArthur, J.M., Howarth, R.J. and Bailey, T.R. (2001) Strontium Isotope Stratigraphy:LOWESS Version 3:Best Fit to the Marine Sr-Isotope Curve for 0–509Ma and Accompanying Look-up Table for Deriving Numerical Age. *Journal of Geology* **109**, 155–170. University of Chicago Press.

Park, G. (2012) *Introducing Tectonics, Rock Structures and Mountain Belts*. Edinburgh: Dunedin Academic Press.

Vail, P.R., Mitchum, R.M. and Thompson, III, S. (1977) Seismic stratigraphy and global changes of sea level, part 3: Relative changes of sea level from coastal onlap. In Payton, C.E. (ed) (1977) AAPG Memoir 26: Seismic stratigraphy—Applications to hydrocarbon exploration: 63–97.

Van Wagoner, J.C., Mitchum, R.M., Campion, K.M., and Rahmanian, V.D. (1990) *Siliciclastic sequence stratigraphy in well logs, cores, and outcrops: Tulsa, Oklahoma*. American Association of Petroleum Geologists Methods in Exploration Series no. 7, 55p.

Van Wagoner, J.C., Posamentier, H.W., Mitchum, R.M., Vail, P.P., Sarg, J.F., Loutit, T.S. and Hordenbol, J. (1988) An overview of the fundamentals of sequence stratigraphy and key definitions. In Wilgus, C.K., Hastings, B.S., Kendall, C.G. St.C., Posamentier, H.W., Ross, C.A. and Van Wagoner, J.C. (eds) (1988) *Sea Level Changes – An Integrated Approach*. SEPM Special Publication, pp39–45.

Willis, Brian and Fitris, Faizil (2012) Sequence Stratigraphy of Miocene Tide-Influenced Sandstones in the Minas Field, Sumatra, Indonesia. *Journal of Sedimentary Research*, 82. 400–421.

Wyse Jackson, P.N. (2010) *Introducing Palaeontology: a Guide to Ancient Life*. Edinburgh: Dunedin Academic Press.

On-line resources

AAPG Methods in Exploration Series no 7

http://www.aapg.org/publications/special-publications/books/pfcatid/876/methods-in-exploration

An Online Guide to Sequence Stratigraphy: University of Georgia Stratigraphy Lab: http://strata.uga.edu/sequence/

International Chronostratigraphic Chart: International Commission on Stratigraphy www.stratigraphy.org/

PALEOMAP Project: www.scotese.com/scotesepubs.htm

SEPM Society for Sedimentary Geology: https://www.sepmstrata.org/Home

Image credits

If not listed here credits are the authors own, the publisher's or Shutterstock.

1.2 Aerial view of the Nile delta. Jacques Descloitres, MODIS Rapid Response Team, NASA/GSFC

1.3 Mississippi delta [Public domain], via Wikimedia Commons

1.4 Deltaic deposits forming in a lake or marine environment, after NASA

1.6 The rock cycle, after www.tes.com

1.7 Varve couplets, after www.tulane.edu

1.8 Varved clays, Austria, by Haneburger [Public Domain], via Wikimedia Commons

1.9 After Science Foundation Course, The Open University.

2.1 Steno's four laws of stratigraphy. After Jones, *Introducing Stratigraphy*, 2015, DAP, Edinburgh

2.2 Stratigraphic ranges and origins of major animal and plant groups. After pubs.usgs.gov/gip/fossils/fig11.gif

2.3 Principle of biostratigraphy. After Jones, *Introducing Stratigraphy*, 2015, DAP, Edinburgh

2.4 First geological map of Great Britain. From: http//www.livescience.com/449-map-changed-world.html

2.5 Age range of fossil species and their use as zone fossils. After http://www.nature.com/scitable/content/ne0000/ne0000/ne0000/ne0000/107977575/Fig.4_1_2.jpg

2.6 Graptolite *Didymograptus murchisoni*. https://upload.wikimedia.org/wikipedia/commons/thumb/7/7e/Didymograptus_murchisoni_lg.jpg. Kevin Walsh, retouched by Verisimilus

2.11 Triassic marine sequence, Utah. https://upload.wikimedia.org/wikipedia/commons/0/08/Triassic_Utah.JPG

2.15 Stratigraphical column. After P. Wyse Jackson, Introducing Palaeontology, 2010, DAP, Edinburgh

3.2 Plate distribution. Adapted from Park, *Introducing Tectonics, Rock Structures and Mountain Belts*, 2012, DAP, Edinburgh

3.3 Convection cells in the Earth's mantle. Adapted from the USGS illustration at http://pubs.usgs.gov/gip/dynamic/unanswered.html

3.4 Types of plate boundaries.After USGS illustration by Jose F. Vigil. http://pubs.usgs.gov/gip/dynamic/Vigil.html

3.5 Destructive plate margin of continent-continent type. https://upload.wikimedia.org/wikipedia/commons/8/83/Continental-continental_convergence_Fig21contcont.gif

3.6 Formation of the Himalayas by collision with India. After USGS illustration at http://pubs.usgs.gov/gip/dynamic/himalaya.html

3.8 Stages of the Wilson Cycle. After Hannes Grobe. http://www.slideshare.net/zombraweb/wilson-cycle

3.10 Site of Chicxulub impact crater, Gulf of Mexico. After NASA Astrophysics Science Division, Goddard Space Flight Center

3.15 Geological succession at the Grand Canyon. After USGS, http://education.usgs.gov/images/schoolyard/GrandCanyonAge.jpg

3.18 Illustration of Walther's Law. Adapted from S. Jones, Introducing Sedimentology, 2015, DAP, Edinburgh

4.1 James Ussher's Annales Veteris Testament (1650). Wing, Early English Books, (1641-1700) [Public Domain] via Wikimedia Commons.

4.2 Atomic structure of hydrogen and its isotopes. https://simple.wikipedia.org/wiki/Isotope#/media/File:Blausen_0530_HydrogenIsotope

4.6 Calibrating the stratigraphic column using radiometric dates. After Earth History 1, Science Foundation Course, The Open University, 1971.

4.7 Stratigraphic column. After Upton, *Volcanoes and the making of Scotland 2015*, DAP, Edinburgh.

4.8 *Cyclomedusa plana*, an Ediacaran fossil from the Flinders Range, Australia. From P. Wyse Jackson, *Introducing Palaeontology*, 2010, DAP, Edinburgh

4.9 The calibrated stratigraphical column. After P. Wyse Jackson, *Introducing Palaeontology*, 2010, DAP, Edinburgh

5.1 The Supercontinent Rodinia. https://commons.wikimedia.org/wiki/File:Rodinia_reconstruction.jpg

5.2 The major glaciations. After Wiki Commons https://commons.wikimedia.org/wiki/File:GlaciationsinEarthExistancelicenced.jpg

5.3 Time line of periods of glaciations in the Phanerozoic https://en.wikipedia.org/wiki/Timeline_of_glaciation

5.6 Approximate extent of the Karoo glaciations. https://cy.wikipedia.org/wiki/Oes_I%C3%A2_Karoo

5.7 Te Nooitgedacht glacial pavement in South Africa. https://en.wikipedia.org/wiki/Nooitgedacht_Glacial_Pavements

5.8 Northern Hemisphere glaciations during the last Glacial Maximum about 18 kya.
Scotese, C. R., 2001. Atlas of Earth History, Volume 1, Paleogeography, PALEOMAP Project, Arlington, Texas, 52 pp.

5.9 Isostatic readjustment of the crust after ice melting. https://kjolurrun.files.wordpress.com/2015/05/rebound.png?w=300&h=208

5.10 Post-glacial rebound effects on the land level of Britain and Ireland. https://upload.wikimedia.org/wikipedia/en/b/b3/Post-glacial_rebound_in_British_Isles.PNG

5.12 Impact craters on the moon. NASA image. https://www.nasa.gov/sites/default/files/thumbnails/image/edu_impact-crater_large.jpg

5.13 Meteor Crater, Arizona. USGS image. https://commons.wikimedia.org/wiki/File:Barringer_Crater_aerial_photo_by_USGS.jpg

5.14 © Graeme Churchard – Wikimedia common links.

5.16 Atmospheric oxygen concentrations over the last 1000myr. From https://upload.wikimedia.org/wikipedia/commons/f/f3/ Sauerstoffgehalt-1000mj2.png

5.18 Time chart of Precambrian time and major stages in Earth history. After http://www.athenapub.com/aria1/mid/md_precambrian1.jpg

5.19 Mass extinction events. After P. Wyse Jackson, Introducing Palaeontology, 2010, DAP, Edinburgh

5.20 Continent reconstruction at 550Myr. Scotese, C. R., 2001. Atlas of Earth History, Volume 1, Paleogeography, PALEOMAP Project, Arlington, Texas, 52 pp.

5.21 Continent reconstruction at 514Myr. Scotese, C. R., 2001. Atlas of Earth History, Volume 1, Paleogeography, PALEOMAP Project, Arlington, Texas, 52 pp.

5.22 Continent reconstruction at 500Myr. Scotese, C. R., 2001. Atlas of Earth History, Volume 1, Paleogeography, PALEOMAP Project, Arlington, Texas, 52 pp.

5.23 Continental reconstruction at 420Myr. After: *The Geology of Northern Ireland- Our Natural Foundation. 2004 Mitchell W.I. (ed). Second Edition. Geological Survey of Northern Ireland.*

5.24 Continent reconstruction at 356Myr. Scotese, C. R., 2001. Atlas of Earth History, Volume 1, Paleogeography, PALEOMAP Project, Arlington, Texas, 52 pp.

5.26 Extent of the Supercontinent Pangaea at 255myr. Scotese, C. R., 2001. Atlas of Earth History, Volume 1, Paleogeography, PALEOMAP Project, Arlington, Texas, 52 pp.

5.27 Palaeogeography at 237myr. Scotese, C. R., 2001. Atlas of Earth History, Volume 1, Paleogeography, PALEOMAP Project, Arlington, Texas, 52 pp.

5.28 Palaeogeography at 152myr. Scotese, C. R., 2001. Atlas of Earth History, Volume 1, Paleogeography, PALEOMAP Project, Arlington, Texas, 52 pp.

5.29 Palaeogeography at 94myr. Scotese, C. R., 2001. Atlas of Earth History, Volume 1, Paleogeography, PALEOMAP Project, Arlington, Texas, 52 pp.

5.30 Cretaceous-Palaeogene boundary at Drumheller,Alberta, Canada. https://en.wikipedia.org/wiki/Cretaceous%E2%80%93Paleogene_boundary#/media/File:KT_boundary_054.jpg

5.32 Palaeogeography at 14myr. Scotese, C. R., 2001. Atlas of Earth History, Volume 1, Paleogeography, PALEOMAP Project, Arlington, Texas, 52 pp.

5.35 Breakup of Supercontinent Pangaea. After USGS: http://pubs.usgs.gov/gip/dynamic/historical.html

5.36 Coccolith skeleton. https://en.wikipedia.org/wiki/Coccolith#/media/File:Emiliania_huxleyi_coccolithophore_(PLoS).png

6.1 International Chronostratigraphic Chart. *Reproduced by permission©ICS International Commission on Stratigraphy [2018].*

6.2 GSSP for the base of the Ediacaran Period, Australia. Photograph courtesy of Dr Alex Liu, University of Cambridge

6.3 Close up of the Golden Spike. Photograph courtesy of Dr Alex Liu, University of Cambridge

6.4 Fish scales and plant debris, Ludlow Bone Bed. Photograph courtesy of the Shrewsbury Museum Service.

6.5 Detail of revised Silurian-Devonian junction, International Chronostratigraphic Chart. *Reproduced by permission ©ICS International Commission on Stratigraphy [2018].*

7.2 Wireline log consisting of a complete set of logs. After: https://upload.wikimedia.org/ wikipedia/commons/5/5e/LI1LOG.jpg By USGS [Public domain], via Wikimedia Commons

7.3 A simplified version of a wireline log. After: http://www.wikiwand.com/en/Petrophysics

7.5 A model for the formation of magnetic striping. After: http://pubs.usgs.gov/gip/dynamic/developing.html

7.6 The Geomagnetic Polarity Time Scale during the last 5myr After: United States Geological Survey Open-File Report 03-187

7.7 The Geomagnetic Polarity Time Scale for the last 169myr. After: https://en.wikipedia.org/wiki/Geomagnetic_reversal#/ media/File:Geomagnetic_polarity_0-169_Ma.svg

7.8 Marine sediment Sr-isotope curve for the Phanerozoic, after McArthur et al, 2001. Adapted from: https://www.sciencedirect.com/science/article/pii/B978044459425900007X

7.9 Seismic data acquisition in a marine setting. Adapted from: https://archive.epa.gov/esd/archive-geophysics/web/html/marine_seismic_methods.html

7.10 Collection of seismic refraction data. Adapted from: http://www.eo-miners.eu/images/ eom_methods/seismic_refraction1.png

7.11 Collection of seismic reflection data. Adapted from: http://www.geologicresources.com/ seismic_reflection500x368.gif

7.12 Seismic reflection data from the Gulf of Mexico. https://woodshole.er.usgs.gov/ projectpages/hydrates/images/image028.jpg

7.13 Basal irregular structures filled by sediments showing original horizontality. After: http://www.geologyin.com/2014/03/principle-of-original-horizontality.html

7.14 Accommodation space available for sediment accumulation. Adapted from: http://www.sepmstrata.org/Terminology.aspx?id=accommodation

7.15 (A) Chronostratigraphic correlation chart or Wheeler diagram. Adapted from: http://www.sepmstrata. org/CMS_Images/ChronoHeader.jpg, courtesy of Dr Chris Kendall. (B) Reflection termination patterns, adapted from Geocities.org

7.16 Changes in sea level through geological time (after Vail *et al.*, 1977). Adapted from: https://www.geol.umd.edu/~jmerck/geol342/images/19slcycles.jpg

7.17 Sequence of events in the four systems tract model. Adapted from Willis, Brian & Fitris, Faizil (2012)

7.18. 7.19, 7.20 Stacking of parasequences. Reprinted from Van Wagoner et al. (1990), with permission from AAPG, whose permission is required for further use: see https://strata.uga.edu/sequence/parasets.html

7.21 The main sequence boundaries of the Four Systems Tract Model for sequence stratigraphy. Based on: http://www.sepmstrata.org/page.aspx?pageid=15, courtesy of Dr Chris Kendall.

7.22 Typical graphic log: Adapted from: https://www.slideshare.net/angelabentley/graphiclogs

7.23 Cross-bedding in sandstones in the canyons of the Escalante region, Utah, USA. https://en.wikipedia.org/wiki/Cross-bedding#/media/File:DryForkDome.jpg

7. 24 Planar cross-bedding and trough cross-bedding. Adapted from http://strata.uga.edu/4500/xstrat1/xstrat1.html

7.25 A parasequence along a sandy, wave- or fluvial-dominated coastline showing a coarsening–upward succession. See https://strata.uga.edu/sequence/parasequences.html Reprinted from Van Wagoner et al. (1990), with permission from AAPG, whose permission is required for further use

7.26 Progradational stacking sequence on a sandy, wave-dominated shore, Utah, USA. See https://strata.uga.edu/sequence/progradationalstacking.html Photograph courtesy of Steven M. Holland.

7.27 A parasequence developed on a muddy siliciclastic shoreline. See https://strata.uga.edu/ sequence/progradationalstacking.html Reprinted from Van Wagoner et al. (1990), with permission from AAPG, whose permission is required for further use

7.28 Tidal Flat Parasequence, in the Ordovician Juniata Formation, West Virginia. https:// strata.uga.edu/sequence/tidalflatparasequence.html Photograph courtesy of Steven M. Holland

7.29 Showing the study area in the Cretaceous Blackhawk Formation, Utah, USA. http://www.sepmstrata.org/CMS_Images/Book-Cliffs-Can-Kenilworth.jpg From: Coe, A.L. *et al.,* The Sedimentary Record of Sea-Level, © 2003 Open University Press, published by Cambridge University Press

7.30 (A) Bookcliffs high frequency clastic parasequence sets (after Coe et al, 2003). Figure 7.30 (B) Graphic log of typical wave dominated shoreface succession showing lithofacies association. Both figures from: Coe, A.L. *et al.,* The Sedimentary Record of Sea-Level, © 2003 Open University Press, published by Cambridge University Press

7.31 Graphic log showing sequence stratigraphy of the Kennilworth Section. See: http://www.sepmstrata.org/page.aspx?pageid=767 Reprinted from Van Wagoner et al. (1990), with permission from AAPG, whose permission is required for further use

7.32 Graphic logs of the sedimentary successions at Panther Canyon, Kennilworth and Coal Canyon. See http://www.sepmstrata.org/page.aspx?pageid=767 Reprinted from Van Wagoner et al. (1990), with permission from AAPG, whose permission is required for further use

7.33 Graphic logs and proposed correlation of the Panther Canyon, Kennilworth and Coal Canyon sections, Book Cliff outcrops, Utah, USA. http://www.sepmstrata.org/CMS_Images/Exer2BWThreeMeas-SectProb.gif Reprinted from Van Wagoner et al. (1990), with permission from AAPG, whose permission is required for further use.

7.35 Conodonts from the Glen Dean Formation. https://commons.wikimedia.org/wiki/File:Conodonts_from_the_Glen_Dean_formation_(Chester)_of_the_Illinois_basin_(1958)_(20654539046).jpg

7.36 Radiolaria from the Challenger Expedition. https://en.wikipedia.org/wiki/Radiolaria#/media/File:Radiolaria.jpg

7.37 Marine diatoms. https://upload.wikimedia.org/wikipedia/commons/2/29/ Diatom2.jpg

7.38 Late Silurian chitinozoan. https://en.wikipedia.org/wiki/Chitinozoan#/media/File:Whole_chitinozoan_cropped.jpg

7.39 Dinoflagellata. https://upload.wikimedia.org/wikipedia/commons/2/2d/Ceratium_hirundinella.jpg